WEST-E / PRAXIS II 0061

Mathematics
Teacher Certification Exam

By: Sharon Wynne, M.S
Southern Connecticut State University

"And, while there's no reason yet to panic, I think it's only prudent that we make preparations to panic."

XAMonline, INC.
Boston

Copyright © 2008 XAMonline, Inc.
All rights reserved. No part of the material protected by this copyright notice may be reproduced or utilized in any form or by any means, electronic or mechanical, including photocopying, recording or by any information storage and retrievable system, without written permission from the copyright holder.

To obtain permission(s) to use the material from this work for any purpose including workshops or seminars, please submit a written request to:

XAMonline, Inc.
21 Orient Ave.
Melrose, MA 02176
Toll Free 1-800-509-4128
Email: info@xamonline.com
Web www.xamonline.com
Fax: 1-781-662-9268

Library of Congress Cataloging-in-Publication Data

Wynne, Sharon A.
 Mathematics 0061: Teacher Certification / Sharon A. Wynne. -2nd ed. ISBN 978-1-58197-674-8
  1. Mathematics 0061.   2. Study Guides.   3. WEST-E
  4. Teachers' Certification & Licensure.   5. Careers

Disclaimer:
The opinions expressed in this publication are the sole works of XAMonline and were created independently from the National Education Association, Educational Testing Service, or any State Department of Education, National Evaluation Systems or other testing affiliates.

Between the time of publication and printing, state specific standards as well as testing formats and website information may change that is not included in part or in whole within this product. Sample test questions are developed by XAMonline and reflect similar content as on real tests; however, they are not former tests. XAMonline assembles content that aligns with state standards but makes no claims nor guarantees teacher candidates a passing score. Numerical scores are determined by testing companies such as NES or ETS and then are compared with individual state standards. A passing score varies from state to state.

Printed in the United States of America

WEST-E/PRAXIS II: Mathematics 0061
ISBN: 978-1-58197-674-8

TEACHER CERTIFICATION STUDY GUIDE

About the Subject Assessments

Purpose: The assessments are designed to test the knowledge and competencies of prospective secondary level teachers. The question bank from which the assessment is drawn is undergoing constant revision. As a result, your test may include questions that will not count towards your score.

Test Version: There are three versions of subject assessment for Mathematics tests in the Praxis Series. Mathematics –Content Knowledge (0061) emphasizes comprehension in Algebra and Number Theory; Measurement, Geometry and Trigonometry; Function and Calculus; Data Analysis, Statistics and Probability; Matrix Algebra and Discrete Mathematics. Mathematics –Proofs, Models and Problems, Part 1 (0063) emphasizes comprehension in Mathematical Problem Solving; Mathematical Reasoning and Proof; Mathematical Connections; Use of Technology. Mathematics –Pedagogy (0065) emphasizes comprehension in Planning Instruction; Implementing Instruction; Assessing Instruction. The Mathematics examination guide is based on a typical knowledge level of persons who have completed a *bachelor's degree program* in Mathematics.

Time Allowance and Format: You will have 1 hour to finish Mathematics – Proofs, Models and Problems, Part 1 (0063) and Mathematics –Pedagogy (0065). You will have 2 hours to finish Mathematics –Content Knowledge (0061). Mathematics –Content Knowledge (0061) consists entirely of multiple-choice questions. In Mathematics –Proofs, Models and Problems, Part 1 (0063) and Mathematics –Pedagogy (0065) consist of extended-response questions. In Mathematics –Content Knowledge (0061) there are approximately 50 multiple-choice questions in the exam. In Mathematics –Proofs, Models and Problems, Part 1 (0063) and Mathematics –Pedagogy (0065) there are approximately 9 extended-response questions in the exam.

Weighting: In the Mathematics –Content Knowledge (0061) exam there are approximately 8 multiple-choice questions in Algebra and Number Theory making 16% of the total content in the exam; 3 multiple-choice questions in Measurement making 6% of the total content in the exam; 5 multiple-choice questions in Geometry making 10% of the total content in the exam; 4 multiple-choice questions in Trigonometry making 8% of the total content in the exam; 8 multiple-choice questions in Function making 16% of the total content in the exam; 6 multiple-choice questions in Calculus making 12% of the total content in the exam; 5-6 multiple-choice questions in Data Analysis and Statistics making 10-12% of the total content in the exam; 2-3 multiple-choice questions in Probability making 2-3% of the total content in the exam; 4-5 multiple-choice questions in Matrix Algebra making 8-10% of the total content in the exam; 3-4 multiple-choice questions in Discrete Mathematics making 6-8% of the total content in the exam.

MATHEMATICS

TEACHER CERTIFICATION STUDY GUIDE

In the Mathematics –Proofs, Models and Problems, Part 1 (0063) exam there are 2 problems representing 40% of the total content in the exam; 1 proof representing 30% of the total content in the exam; 1 model representing 30% of the total content in the exam, all consisting of allotted topical content for Mathematics

Additional Information about the PRAXIS Assessments: The PRAXIS series subject assessments are developed *Educational Testing Service.* They provide additional information on the PRAXIS series assessments, including registration, preparation and testing procedures, study materials such as three topical guides, one for each subtest, that are all together about 50 pages of information including approximately 41 additional sample questions.

MATHEMATICS

TEACHER CERTIFICATION STUDY GUIDE

| SUBAREA | TABLE OF CONTENTS | PG # |

MATHEMATICS – CONTENT KNOWLEDGE

0001 ALGEBRA AND NUMBER THEORY 1

0002 MEASUREMENT 28

0003 GEOMETRY 36

0004 TRIGONOMETRY 75

0005 FUNCTIONS 91

0006 CALCULUS 104

0007 DATA ANALYSIS, STATISTICS AND PROBABILITY 134

0008 MATRIX ALGEBRA 142

0009 DISCRETE MATHEMATICS 158

MATHEMATICS – PROOFS, MODELS AND PROBLEMS, PART 1

0010 PROBLEMS 171

0011 MODELS 174

0012 PROOFS 176

CURRICULUM AND INSTRUCTION 177

ANSWER KEY TO PRACTICE PROBLEMS 186

SAMPLE TEST 190

ANSWER KEY 205

RIGOR TABLE 206

RATIONALES WITH SAMPLE QUESTIONS 208

TEACHER CERTIFICATION STUDY GUIDE

Great Study and Testing Tips!

What to study in order to prepare for the subject assessments is the focus of this study guide but equally important is *how* you study.

You can increase your chances of truly mastering the information by taking some simple, but effective steps.

Study Tips:

1. Some foods aid the learning process. Foods such as milk, nuts, seeds, rice, and oats help your study efforts by releasing natural memory enhancers called CCKs (*cholecystokinin*) composed of *tryptophan*, *choline*, and *phenylalanine*. All of these chemicals enhance the neurotransmitters associated with memory. Before studying, try a light, protein-rich meal of eggs, turkey, and fish. All of these foods release the memory enhancing chemicals. The better the connections, the more you comprehend.

Likewise, before you take a test, stick to a light snack of energy boosting and relaxing foods. A glass of milk, a piece of fruit, or some peanuts all release various memory-boosting chemicals and help you to relax and focus on the subject at hand.

2. Learn to take great notes. A by-product of our modern culture is that we have grown accustomed to getting our information in short doses (i.e. TV news sound bites or USA Today style newspaper articles.)

Consequently, we've subconsciously trained ourselves to assimilate information better in neat little packages. If your notes are scrawled all over the paper, it fragments the flow of the information. Strive for clarity. Newspapers use a standard format to achieve clarity. Your notes can be much clearer through use of proper formatting. A very effective format is called the *"Cornell Method."*

> Take a sheet of loose-leaf lined notebook paper and draw a line all the way down the paper about 1-2" from the left-hand edge.
>
> Draw another line across the width of the paper about 1-2" up from the bottom. Repeat this process on the reverse side of the page.

Look at the highly effective result. You have ample room for notes, a left hand margin for special emphasis items or inserting supplementary data from the textbook, a large area at the bottom for a brief summary, and a little rectangular space for just about anything you want.

MATHEMATICS

3. Get the concept then the details. Too often we focus on the details and don't gather an understanding of the concept. However, if you simply memorize only dates, places, or names, you may well miss the whole point of the subject.

A key way to understand things is to put them in your own words. If you are working from a textbook, automatically summarize each paragraph in your mind. If you are outlining text, don't simply copy the author's words.

Rephrase them in your own words. You remember your own thoughts and words much better than someone else's, and subconsciously tend to associate the important details to the core concepts.

4. Ask Why? Pull apart written material paragraph by paragraph and don't forget the captions under the illustrations.

Example: If the heading is "Stream Erosion", flip it around to read "Why do streams erode?" Then answer the questions.

If you train your mind to think in a series of questions and answers, not only will you learn more, but it also helps to lessen the test anxiety because you are used to answering questions.

5. Read for reinforcement and future needs. Even if you only have 10 minutes, put your notes or a book in your hand. Your mind is similar to a computer; you have to input data in order to have it processed. *By reading, you are creating the neural connections for future retrieval.* The more times you read something, the more you reinforce the learning of ideas.

Even if you don't fully understand something on the first pass, *your mind stores much of the material for later recall.*

6. Relax to learn so go into exile. Our bodies respond to an inner clock called biorhythms. Burning the midnight oil works well for some people, but not everyone.

If possible, set aside a particular place to study that is free of distractions. Shut off the television, cell phone, pager and exile your friends and family during your study period.

If you really are bothered by silence, try background music. Light classical music at a low volume has been shown to aid in concentration over other types. Music that evokes pleasant emotions without lyrics are highly suggested. Try just about anything by Mozart. It relaxes you.

MATHEMATICS

TEACHER CERTIFICATION STUDY GUIDE

7. Use arrows not highlighters. At best, it's difficult to read a page full of yellow, pink, blue, and green streaks. Try staring at a neon sign for a while and you'll soon see that the horde of colors obscure the message.

A quick note, a brief dash of color, an underline, and an arrow pointing to a particular passage is much clearer than a horde of highlighted words.

8. Budget your study time. Although you shouldn't ignore any of the material, *allocate your available study time in the same ratio that topics may appear on the test.*

Testing Tips:

1. Get smart, play dumb. Don't read anything into the question. Don't make an assumption that the test writer is looking for something else than what is asked. Stick to the question as written and don't read extra things into it.

2. Read the question and all the choices *twice* before answering the question. You may miss something by not carefully reading, and then re-reading both the question and the answers.

If you really don't have a clue as to the right answer, leave it blank on the first time through. Go on to the other questions, as they may provide a clue as to how to answer the skipped questions.

If later on, you still can't answer the skipped ones . . . **Guess.** The only penalty for guessing is that you *might* get it wrong. Only one thing is certain; if you don't put anything down, you will get it wrong!

3. Turn the question into a statement. Look at the way the questions are worded. The syntax of the question usually provides a clue. Does it seem more familiar as a statement rather than as a question? Does it sound strange?

By turning a question into a statement, you may be able to spot if an answer sounds right, and it may also trigger memories of material you have read.

4. Look for hidden clues. It's actually very difficult to compose multiple-foil (choice) questions without giving away part of the answer in the options presented.

In most multiple-choice questions you can often readily eliminate one or two of the potential answers. This leaves you with only two real possibilities and automatically your odds go to Fifty-Fifty for very little work.

5. Trust your instincts. For every fact that you have read, you subconsciously retain something of that knowledge. On questions that you aren't really certain about, go with your basic instincts. **Your first impression on how to answer a question is usually correct.**

6. Mark your answers directly on the test booklet. Don't bother trying to fill in the optical scan sheet on the first pass through the test.

Just be very careful not to miss-mark your answers when you eventually transcribe them to the scan sheet.

7. Watch the clock! You have a set amount of time to answer the questions. Don't get bogged down trying to answer a single question at the expense of 10 questions you can more readily answer.

MATHEMATICS

THIS PAGE BLANK

0001 ALGEBRA AND NUMBER THEORY

To convert a fraction to a decimal, simply divide the numerator (top) by the denominator (bottom). Use long division if necessary.

If a decimal has a fixed number of digits, the decimal is said to be terminating. To write such a decimal as a fraction, first determine what place value the farthest right digit is in, for example: tenths, hundredths, thousandths, ten thousandths, hundred thousands, etc. Then drop the decimal and place the string of digits over the number given by the place value.

If a decimal continues forever by repeating a string of digits, the decimal is said to be repeating. To write a repeating decimal as a fraction, follow these steps.

a. Let x = the repeating decimal
(ex. $x = .716716716...$)
b. Multiply x by the multiple of ten that will move the decimal just to the right of the repeating block of digits.
(ex. $1000x = 716.716716...$)
c. Subtract the first equation from the second.
(ex. $1000x - x = 716.716.716... - .716716...$)
d. Simplify and solve this equation. The repeating block of digits will subtract out.
(ex. $999x = 716$ so $x = {}^{716}\!/_{999}$)
e. The solution will be the fraction for the repeating decimal.

The real number properties are best explained in terms of a small set of numbers. For each property, a given set will be provided.

Axioms of Addition

Closure—For all real numbers a and b, $a + b$ is a unique real number.

Associative—For all real numbers a, b, and c, $(a + b) + c = a + (b + c)$.

Additive Identity—There exists a unique real number 0 (zero) such that $a + 0 = 0 + a = a$ for every real number a.

Additive Inverses—For each real number a, there exists a real number $-a$ (the opposite of a) such that $a + (-a) = (-a) + a = 0$.

Commutative—For all real numbers a and b, $a + b = b + a$.

Axioms of Multiplication

Closure—For all real numbers a and b, ab is a unique real number.

Associative—For all real numbers a, b, and c, $(ab)c = a(bc)$.

Multiplicative Identity—There exists a unique nonzero real number 1 (one) such that $1 \cdot a = a \cdot 1 = a$.

Multiplicative Inverses—For each nonzero real number, there exists a real number $1/a$ (the reciprocal of a) such that $a(1/a) = (1/a)a = 1$.

Commutative—For all real numbers a and b, $ab = ba$.

The Distributive Axiom of Multiplication over Addition

For all real numbers a, b, and c, $a(b + c) = ab + ac$.

a. **Natural numbers**--the counting numbers, 1,2,3,...

b. **Whole numbers**--the counting numbers along with zero, 0,1,2...

c. **Integers**--the counting numbers, their opposites, and zero, ..., ⁻1,0,1,...

d. **Rationals**--all of the fractions that can be formed from the whole numbers. Zero cannot be the denominator. In decimal form, these numbers will either be terminating or repeating decimals. Simplify square roots to determine if the number can be written as a fraction.

e. **Irrationals**--real numbers that cannot be written as a fraction. The decimal forms of these numbers are neither terminating nor repeating. Examples: $\pi, e, \sqrt{2}$, etc.

f. **Real numbers**--the set of numbers obtained by combining the rationals and irrationals. Complex numbers, i.e. numbers that involve i or $\sqrt{-1}$, are not real numbers.

The **Denseness Property** of real numbers states that, if all real numbers are ordered from least to greatest on a number line, there is an infinite set of real numbers between any two given numbers on the line.

Example:

Between 7.6 and 7.7, there is the rational number 7.65 in the set of real numbers.

Between 3 and 4 there exists no other natural number.

For standardization purposes, there is an accepted order in which operations are performed in any given algebraic expression. The following mnemonic is often used for the order in which operations are performed.

Please	Parentheses	
Excuse	Exponents	
My	Multiply	Multiply or Divide depending on which
Dear	Divide	operation is encountered first from left to right.
Aunt	Add	Add or Subtract depending on which
Sally	Subtract	operation is encountered first from left to right.

Subtraction is the inverse of Addition, and vice-versa.
Division is the inverse of Multiplication, and vice-versa.
Taking a square root is the inverse of squaring, and vice-versa.

These inverse operations are used when solving equations.

Prime numbers are numbers that can only be factored into 1 and the number itself. When factoring into prime factors, all the factors must be numbers that cannot be factored again (without using 1). Initially numbers can be factored into any 2 factors. Check each resulting factor to see if it can be factored again. Continue factoring until all remaining factors are prime. This is the list of prime factors. Regardless of what way the original number was factored, the final list of prime factors will always be the same.

Example: Factor 30 into prime factors.

Factor 30 into any 2 factors.

5 · 6	Now factor the 6.
5 · 2 · 3	These are all prime factors.

Factor 30 into any 2 factors.

3 · 10	Now factor the 10.
3 · 2 · 5	These are the same prime factors even though the original factors were different.

Example: Factor 240 into prime factors.

Factor 240 into any 2 factors.

24 · 10	Now factor both 24 and 10.
4 · 6 · 2 · 5	Now factor both 4 and 6.
2 · 2 · 2 · 3 · 2 · 5	These are prime factors.

This can also be written as $2^4 \cdot 3 \cdot 5$.

Divisibility Tests and Divisors

a. A number is divisible by 2 if that number is an even number (which means it ends in 0,2,4,6 or 8).

1,354 ends in 4, so it is divisible by 2. 240,685 ends in a 5, so it is not divisible by 2.

b. A number is divisible by 3 if the sum of its digits is evenly divisible by 3.

The sum of the digits of 964 is 9+6+4 = 19. Since 19 is not divisible by 3, neither is 964. The digits of 86,514 is 8+6+5+1+4 = 24. Since 24 is divisible by 3, 86,514 is also divisible by 3.

c. A number is divisible by 4 if the number in its last 2 digits is evenly divisible by 4.

The number 113,336 ends with the number 36 in the last 2 columns. Since 36 is divisible by 4, then 113,336 is also divisible by 4.

The number 135,627 ends with the number 27 in the last 2 columns. Since 27 is not evenly divisible by 4, then 135,627 is also not divisible by 4.

d. A number is divisible by 5 if the number ends in either a 5 or a 0.

225 ends with a 5 so it is divisible by 5. The number 470 is also divisible by 5 because its last digit is a 0. 2,358 is not divisible by 5 because its last digit is an 8, not a 5 or a 0.

e. A number is divisible by 6 if the number is even and the sum of its digits is evenly divisible by 3.

4,950 is an even number and its digits add to 18. (4+9+5+0 = 18) Since the number is even and the sum of its digits is 18 (which is divisible by 3), then 4950 is divisible by 6. 326 is an even number, but its digits add up to 11. Since 11 is not divisible by 3, then 326 is not divisible by 6. 698,135 is not an even number, so it cannot possibly be divided evenly by 6.

f. A number is divisible by 8 if the number in its last 3 digits is evenly divisible by 8.

The number 113,336 ends with the 3-digit number 336 in the last 3 places. Since 336 is divisible by 8, then 113,336 is also divisible by 8.

The number 465,627 ends with the number 627 in the last 3 places. Since 627 is not evenly divisible by 8, then 465,627 is also not divisible by 8.

g. A number is divisible by 9 if the sum of its digits is evenly divisible by 9.

The sum of the digits of 874 is 8+7+4 = 19. Since 19 is not divisible by 9, neither is 874. The digits of 116,514 is 1+1+6+5+1+4 = 18. Since 18 is divisible by 9, 116,514 is also divisible by 9.

h. A number is divisible by 10 if the number ends in the digit 0.

305 ends with a 5 so it is not divisible by 10. The number 2,030,270 is divisible by 10 because its last digit is a 0. 42,978 is not divisible by 10 because its last digit is an 8, not a 0.

i. Why these rules work.

All even numbers are divisible by 2 by definition. A 2-digit number (with T as the tens digit and U as the ones digit) has as its sum of the digits, T + U. Suppose this sum of T + U is divisible by 3. Then it equals 3 times some constant, K. So, T + U = 3K. Solving this for U, U = 3K - T. The original 2 digit number would be represented by 10T + U. Substituting 3K - T in place of U, this 2-digit number becomes 10T + U = 10T + (3K - T) = 9T + 3K. This 2-digit number is clearly divisible by 3, since each term is divisible by 3. Therefore, if the sum of the digits of a number is divisible by 3, then the number itself is also divisible by 3. Since 4 divides evenly into 100, 200, or 300, 4 will divide evenly into any amount of hundreds. The only part of a number that determines if 4 will divide into it evenly is the number in the last 2 places. Numbers divisible by 5 end in 5 or 0. This is clear if you look at the answers to the multiplication table for 5. Answers to the multiplication table for 6 are all even numbers. Since 6 factors into 2 times 3, the divisibility rules for 2 and 3 must both work. Any number of thousands is divisible by 8. Only the last 3 places of the number determine whether or not it is divisible by 8. A 2 digit number (with T as the tens digit and U as the ones digit) has as its sum of the digits, T + U.

Suppose this sum of T + U is divisible by 9. Then it equals 9 times some constant, K. So, T + U = 9K. Solving this for U, U = 9K - T. The original 2-digit number would be represented by 10T + U. Substituting 9K - T in place of U, this 2-digit number becomes 10T + U = 10T + (9K - T) = 9T + 9K. This 2-digit number is clearly divisible by 9, since each term is divisible by 9. Therefore, if the sum of the digits of a number is divisible by 9, then the number itself is also divisible by 9. Numbers divisible by 10 must be multiples of 10 which all end in a zero.

Prime numbers are whole numbers greater than 1 that have only 2 factors, 1 and the number itself. Examples of prime numbers are 2,3,5,7,11,13,17, or 19. Note that 2 is the only even prime number.

Composite numbers are whole numbers that have more than 2 different factors. For example 9 is composite because besides factors of 1 and 9, 3 is also a factor. 70 is also composite because besides the factors of 1 and 70, the numbers 2,5,7,10,14, and 35 are also all factors.

Remember that the number 1 is neither prime nor composite.

The **exponent form** is a shortcut method to write repeated multiplication. The **base** is the factor. The **exponent** tells how many times that number is multiplied by itself.

The following are basic rules for exponents:

$a^1 = a$ for all values of a; thus $17^1 = 17$

$b^0 = 1$ for all values of b; thus $24^0 = 1$

$10^n = 1$ with n zeros; thus $10^6 = 1,000,000$

- In order to add or subtract rational expressions, they must have a common denominator. If they don't have a common denominator, then factor the denominators to determine what factors are missing from each denominator to make the LCD.

Multiply both numerator and denominator by the missing factor(s). Once the fractions have a common denominator, add or subtract their numerators, but keep the common denominator the same. Factor the numerator if possible and reduce if there are any factors that can be cancelled.

1. Find the least common denominator for $6a^3b^2$ and $4ab^3$.

   These factor into $2 \cdot 3 \cdot a^3 \cdot b^2$ and $2 \cdot 2 \cdot a \cdot b^3$.
   The first expression must be multiplied by another 2 and b.
   The other expression must be multiplied by 3 and a^2.
   Then both expressions would be
   $2 \cdot 2 \cdot 3 \cdot a^3 \cdot b^3 = 12a^3b^3 = $ LCD.

2. Find the LCD for $x^2 - 4$, $x^2 + 5x + 6$, and $x^2 + x - 6$.

   $x^2 - 4 \quad$ factors into $(x-2)(x+2)$
   $x^2 + 5x + 6 \quad$ factors into $(x+3)(x+2)$
   $x^2 + x - 6 \quad$ factors into $(x+3)(x-2)$

To make these lists of factors the same, they must all be $(x+3)(x+2)(x-2)$. This is the LCD.

3.

$$\frac{5}{6a^3b^2} + \frac{1}{4ab^3} = \frac{5(2b)}{6a^3b^2(2b)} + \frac{1(3a^2)}{4ab^3(3a^2)} = \frac{10b}{12a^3b^3} + \frac{3a^2}{12a^3b^3} = \frac{10b + 3a^2}{12a^3b^3}$$

This will not reduce as all 3 terms are not divisible by anything.

4.
$$\frac{2}{x^2-4} - \frac{3}{x^2+5x+6} + \frac{7}{x^2+x-6} =$$

$$\frac{2}{(x-2)(x+2)} - \frac{3}{(x+3)(x+2)} + \frac{7}{(x+3)(x-2)} =$$

$$\frac{2(x+3)}{(x-2)(x+2)(x+3)} - \frac{3(x-2)}{(x+3)(x+2)(x-2)} + \frac{7(x+2)}{(x+3)(x-2)(x+2)} =$$

$$\frac{2x+6}{(x-2)(x+2)(x+3)} - \frac{3x-6}{(x+3)(x+2)(x-2)} + \frac{7x+14}{(x+3)(x-2)(x+2)} =$$

$$\frac{2x+6-(3x-6)+7x+14}{(x+3)(x-2)(x+2)} = \frac{6x+26}{(x+3)(x-2)(x+2)}$$

This will not reduce.

Try These:

1. $\dfrac{6}{x-3} + \dfrac{2}{x+7}$

2. $\dfrac{5}{4a^2b^5} + \dfrac{3}{5a^4b^3}$

3. $\dfrac{x+3}{x^2-25} + \dfrac{x-6}{x^2-2x-15}$

- Some problems can be solved using equations with rational expressions. First write the equation. To solve it, multiply each term by the LCD of all fractions. This will cancel out all of the denominators and give an equivalent algebraic equation that can be solved.

1. The denominator of a fraction is two less than three times the numerator. If 3 is added to both the numerator and denominator, the new fraction equals $1/2$.

original fraction: $\dfrac{x}{3x-2}$ revised fraction: $\dfrac{x+3}{3x+1}$

$$\dfrac{x+3}{3x+1} = \dfrac{1}{2} \qquad 2x+6 = 3x+1$$
$$x = 5$$

original fraction: $\dfrac{5}{13}$

2. Elly Mae can feed the animals in 15 minutes. Jethro can feed them in 10 minutes. How long will it take them if they work together?

Solution: If Elly Mae can feed the animals in 15 minutes, then she could feed $1/15$ of them in 1 minute, $2/15$ of them in 2 minutes, $x/15$ of them in x minutes. In the same fashion Jethro could feed $x/10$ of them in x minutes. Together they complete 1 job. The equation is:

$$\dfrac{x}{15} + \dfrac{x}{10} = 1$$

Multiply each term by the LCD of 30:

$$2x + 3x = 30$$
$$x = 6 \text{ minutes}$$

3. A salesman drove 480 miles from Pittsburgh to Hartford. The next day he returned the same distance to Pittsburgh in half an hour less time than his original trip took, because he increased his average speed by 4 mph. Find his original speed.

Since distance = rate x time, then time = $\dfrac{\text{distance}}{\text{rate}}$

original time $- 1/2$ hour = shorter return time

$$\dfrac{480}{x} - \dfrac{1}{2} = \dfrac{480}{x+4}$$

Multiplying by the LCD of $2x(x+4)$, the equation becomes:

$480[2(x+4)] - 1[x(x+4)] = 480(2x)$

$960x + 3840 - x^2 - 4x = 960x$

$x^2 + 4x - 3840 = 0$

$(x+64)(x-60) = 0$ Either (x-60=0) or (x+64=0) or both=0

$x = 60$ 60 mph is the original speed.

                This is the solution since the time

$x + 4 = 64$ cannot be negative. Check your answer

$\dfrac{480}{60} - \dfrac{1}{2} = \dfrac{480}{64}$

$8 - \dfrac{1}{2} = 7\dfrac{1}{2}$

$7\dfrac{1}{2} = 7\dfrac{1}{2}$

Try these:

1. Working together, Larry, Moe, and Curly can paint an elephant in 3 minutes. Working alone, it would take Larry 10 minutes or Moe 6 minutes to paint the elephant. How long would it take Curly to paint the elephant if he worked alone?

2. The denominator of a fraction is 5 more than twice the numerator. If the numerator is doubled, and the denominator is increased by 5, the new fraction is equal to 1/2. Find the original number.

3. A trip from Augusta, Maine to Galveston, Texas is 2108 miles. If a car drove 6 mph faster than a truck and got to Galveston 3 hours before the truck, find the speeds of the car and truck.

- To solve an algebraic formula for some variable, called R, follow the following steps:

a. Eliminate any parentheses using the distributive property.
b. Multiply every term by the LCD of any fractions to write an equivalent equation without any fractions.
c. Move all terms containing the variable, R, to one side of the equation. Move all terms without the variable to the opposite side of the equation.
d. If there are 2 or more terms containing the variable R, factor **only R** out of each of those terms as a common factor.
e. Divide both sides of the equation by the number or expression being multiplied times the variable, R.
f. Reduce fractions if possible.
g. Remember there are restrictions on values allowed for variables because the denominator can not equal zero.

1. Solve $A = p + prt$ for t.

$$A - p = prt$$
$$\frac{A-p}{pr} = \frac{prt}{pr}$$
$$\frac{A-p}{pr} = t$$

2. Solve $A = p + prt$ for p.

$$A = p(1+rt)$$
$$\frac{A}{1+rt} = \frac{p(1+rt)}{1+rt}$$
$$\frac{A}{1+rt} = p$$

3. $A = 1/2\, h(b_1 + b_2)$ for b_2

$$A = 1/2\, hb_1 + 1/2\, hb_2 \quad \leftarrow \text{step a}$$
$$2A = hb_1 + hb_2 \quad \leftarrow \text{step b}$$
$$2A - hb_1 = hb_2 \quad \leftarrow \text{step c}$$
$$\frac{2A - hb_1}{h} = \frac{hb_2}{h} \quad \leftarrow \text{step d}$$
$$\frac{2A - hb_1}{h} = b_2 \quad \leftarrow \text{will not reduce}$$

Solve:
1. $F = 9/5\, C + 32$ for C
2. $A = 1/2\, bh + h^2$ for b
3. $S = 180(n-2)$ for n

To graph an inequality, solve the inequality for y. This gets the inequality in **slope intercept form**, (for example : $y < mx + b$). The point (0,b) is the y-intercept and m is the line's slope.

- If the inequality solves to $x \geq$ **any number**, then the graph includes a **vertical line**.

- If the inequality solves to $y \leq$ **any number**, then the graph includes a **horizontal line**.

- When graphing a linear inequality, the line will be dotted if the inequality sign is $<$ or $>$. If the inequality signs are either $\geq$ or $\leq$, the line on the graph will be a solid line. Shade above the line when the inequality sign is $\geq$ or $>$. Shade below the line when the inequality sign is $<$ or $\leq$. For inequalities of the forms $x >$ number, $x \leq$ number, $x <$ number, or $x \geq$ number, draw a vertical line (solid or dotted). Shade to the right for $>$ or $\geq$. Shade to the left for $<$ or $\leq$.

> Remember: **Dividing or multiplying by a negative number will reverse the direction of the inequality sign.**

Use these rules to graph and shade each inequality. The solution to a system of linear inequalities consists of the part of the graph that is shaded for each inequality. For instance, if the graph of one inequality was shaded with red, and the graph of another inequality was shaded with blue, then the overlapping area would be shaded purple. The purple area would be the points in the solution set of this system.

Example: Solve by graphing:

$x + y \leq 6$
$x - 2y \leq 6$

Solving the inequalities for y, they become:

$y \leq {}^-x + 6$ (y intercept of 6 and slope = $^-1$)
$y \geq 1/2x - 3$ (y intercept of $^-3$ and slope = $1/2$)

A graph with shading is shown below:

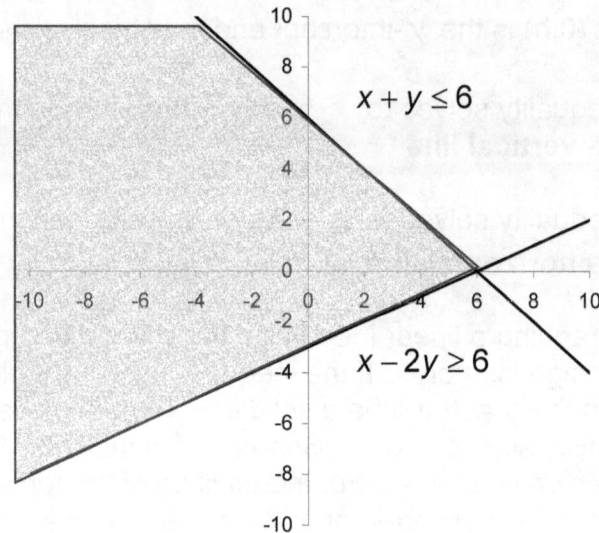

To solve an **equation or inequality**, follow these steps:

STEP 1. If there are parentheses, use the distributive property to eliminate them.

STEP 2. If there are fractions, determine their LCD (least common denominator). Multiply every term of the equation by the LCD. This will cancel out all of the fractions while solving the equation or inequality.

STEP 3. If there are decimals, find the largest decimal. Multiply each term by a power of 10(10, 100, 1000,etc.) with the same number of zeros as the length of the decimal. This will eliminate all decimals while solving the equation or inequality.

STEP 4. Combine like terms on each side of the equation or inequality.

STEP 5. If there are variables on both sides of the equation, add or subtract one of those variable terms to move it to the other side. Combine like terms.

STEP 6. If there are constants on both sides, add or subtract one of those constants to move it to the other side. Combine like terms.

STEP 7. If there is a coefficient in front of the variable, divide both sides by this number. This is the answer to an equation. However, remember:

Dividing or multiplying an inequality by a negative number will reverse the direction of the inequality sign.

STEP 8. The solution of a linear equation solves to one single number. The solution of an inequality is always stated including the inequality sign.

Example: Solve: $3(2x+5)-4x=5(x+9)$

$6x+15-4x=5x+45$ ref. step 1
$2x+15=5x+45$ ref. step 4
$^-3x+15=45$ ref. step 5
$^-3x=30$ ref. step 6
$x=\,^-10$ ref. step 7

Example: Solve: $1/2(5x+34)=1/4(3x-5)$

$5/2\,x+17=3/4\,x-5/4$ ref. step 1
LCD of 5/2, 3/4, and 5/4 is 4.
Multiply by the LCD of 4.
$4(5/2\,x+17)=(3/4\,x-5/4)4$ ref. step 2
$10x+68=3x-5$
$7x+68=\,^-5$ ref. step 5
$7x=\,^-73$ ref. step 6
$x=\,^-73/7$ or $^-10\,3/7$ ref. step 7

Check:

$$\frac{1}{2}\left[5\frac{-13}{7}+34\right]=\frac{1}{4}\left[3\left(\frac{-13}{7}\right)-\frac{5}{4}\right]$$

$$\frac{1}{2}\left[\frac{-13(5)}{7}+34\right]=\frac{1}{4}\left[3\left(\frac{-13}{7}\right)-\frac{5}{4}\right]$$

$$\frac{-13(5)}{7}+17=\frac{3(-13)}{28}-\frac{5}{4}$$

$$\frac{-13(5)+17(14)}{14}=\frac{3(-13)}{28}-\frac{5}{4}$$

$$[-13(5)+17(14)]2=\frac{3(-13)-35}{28}$$

$$\frac{-130+476}{28}=\frac{-219-35}{28}$$

$$\frac{-254}{28}=\frac{-254}{28}$$

Example: Solve: $6x+21<8x+31$

$^-2x+21<31$ ref. step 5

$^-2x<10$ ref. step 6

$x>{}^-5$ ref. step 7

Note that the inequality sign has changed.

To solve an **absolute value equation,** follow these steps:

1. Get the absolute value expression alone on one side of the equation.

2. Split the absolute value equation into 2 separate equations without absolute value bars. Write the expression inside the absolute value bars (without the bars) equal to the expression on the other side of the equation. Now write the expression inside the absolute value bars equal to the opposite of the expression on the other side of the equation.

3. Now solve each of these equations.

4. **Check each answer by substituting them into the original equation** (with the absolute value symbol). There will be answers that do not check in the original equation. These answers are discarded as they are **extraneous solutions**. If all answers are discarded as incorrect, then the answer to the equation is ∅, which means the empty set or the null set. (0, 1, or 2 solutions could be correct.)

To solve an **absolute value inequality**, follow these steps:

1. Get the absolute value expression alone on one side of the inequality. Remember: **Dividing or multiplying by a negative number will reverse the direction of the inequality sign.**

2. Remember what the inequality sign is at this point.

3. Split the absolute value inequality into 2 separate inequalities without absolute value bars. First rewrite the inequality without the absolute bars and solve it. Next write the expression inside the absolute value bar followed by the opposite inequality sign and then by the opposite of the expression on the other side of the inequality. Now solve it.

4. If the sign in the inequality on step 2 is $<$ or $\leq$, the answer is those 2 inequalities connected by the word **and**. The solution set consists of the points between the 2 numbers on the number line. If the sign in the inequality on step 2 is $>$ or $\geq$, the answer is those 2 inequalities connected by the word **or**. The solution set consists of the points outside the 2 numbers on the number line.

   If an expression inside an absolute value bar is compared to a negative number, the answer can also be either all real numbers or the empty set (∅). For instance,

   $$|x+3| < {}^-6$$

   would have the empty set as the answer, since an absolute value is always positive and will never be less than $^-6$. However,

   $$|x+3| > {}^-6$$

   would have all real numbers as the answer, since an absolute value is always positive or at least zero, and will never be less than -6. In similar fashion,

   $$|x+3| = {}^-6$$

   would never check because an absolute value will never give a negative value.

Example: Solve and check:

$$|2x-5|+1=12$$
$$|2x-5|=11 \quad \text{Get absolute value alone.}$$

Rewrite as 2 equations and solve separately.

same equation without absolute value		same equation without absolute value but right side is opposite
$2x-5=11$		$2x-5={}^-11$
$2x=16$	and	$2x={}^-6$
$x=8$		$x={}^-3$

Checks:
$$|2x-5|+1=12 \qquad |2x-5|+1=12$$
$$|2(8)-5|+1=12 \qquad |2({}^-3)-5|+1=12$$
$$|11|+1=12 \qquad |{}^-11|+1=12$$
$$12=12 \qquad 12=12$$

This time both 8 and $^-3$ check.

Example: Solve and check:

$$2|x-7|-13 \geq 11$$

$$2|x-7| \geq 24 \quad \text{Get absolute value alone.}$$

$$|x-7| \geq 12$$

Rewrite as 2 inequalities and solve separately.

same inequality without absolute value		same inequality without absolute value but right side and inequality sign are both the opposite
$x-7 \geq 12$	or	$x-7 \leq {}^-12$
$x \geq 19$	or	$x \leq {}^-5$

Equations and inequalities can be used to solve various types of word problems. Examples follow.

Example: The YMCA wants to sell raffle tickets to raise at least $32,000. If they must pay $7,250 in expenses and prizes out of the money collected from the tickets, how many tickets worth $25 each must they sell?

Solution: Since they want to raise **at least $32,000**, that means they would be happy to get 32,000 **or more**. This requires an inequality.

Let x = number of tickets sold
Then $25x$ = total money collected for x tickets

Total money minus expenses is greater than $32,000.

$$25x - 7250 \geq 32000$$
$$25x \geq 39250$$
$$x \geq 1570$$

If they sell **1,570 tickets or more**, they will raise AT LEAST $32,000.

Example: The Simpsons went out for dinner. All 4 of them ordered the aardvark steak dinner. Bert paid for the 4 meals and included a tip of $12 for a total of $84.60. How much was an aardvark steak dinner?

Let x = the price of one aardvark dinner.
So $4x$ = the price of 4 aardvark dinners.

Let x = the price of one aardvark dinner
So $4x$ = the price of 4 aardavark dinners

$$4x = 84.60 - 12$$
$$4x = 72.60$$
$$x = \frac{72.60}{4} = \$18.15 \quad \text{The price of one aardvark dinner.}$$

Some word problems can be solved using a system of equations or inequalities. Watch for words like greater than, less than, at least, or no more than which indicate the need for inequalities.

Example: Farmer Greenjeans bought 4 cows and 6 sheep for $1700. Mr. Ziffel bought 3 cows and 12 sheep for $2400. If all the cows were the same price and all the sheep were another price, find the price charged for a cow or for a sheep.

Let x = price of a cow
Let y = price of a sheep

Then Farmer Greenjeans' equation would be: $4x + 6y = 1700$
Mr. Ziffel's equation would be: $3x + 12y = 2400$

To solve by **addition-subtraction**:

Multiply the first equation by $^-2$: $\quad ^-2(4x + 6y = 1700)$
Keep the other equation the same: $\quad (3x + 12y = 2400)$
By doing this, the equations can be added to each other to eliminate one variable and solve for the other variable.

$$^-8x - 12y = {}^-3400$$
$$\underline{3x + 12y = 2400} \qquad \text{Add these equations.}$$
$$^-5x \qquad = {}^-1000$$

$\qquad\qquad x = 200 \leftarrow$ the price of a cow was $200.
Solving for y, $y = 150 \leftarrow$ the price of a sheep, $150.

(This problem can also be solved by substitution or determinants.)

Example: John has 9 coins, which are either dimes or nickels, that are worth $.65. Determine how many of each coin he has.

Let d = number of dimes.
Let n = number of nickels.
The number of coins total 9.
The value of the coins equals 65.

Then: $n + d = 9$
$5n + 10d = 65$

Multiplying the first equation by $^-5$, it becomes:

$^-5n - 5d = {}^-45$
$\underline{5n + 10d = 65}$
$5d = 20$

$d = 4$ There are 4 dimes, so there are
$(9-4)$ or 5 nickels.

Example: Sharon's Bike Shoppe can assemble a 3 speed bike in 30 minutes or a 10 speed bike in 60 minutes. The profit on each bike sold is $60 for a 3 speed or $75 for a 10 speed bike. How many of each type of bike should they assemble during an 8 hour day (480 minutes) to make the maximum profit? Total daily profit must be at least $300.

Let x = number of 3 speed bikes.
y = number of 10 speed bikes.

Since there are only 480 minutes to use each day,

$30x + 60y \leq 480$ is the first inequality.
$30x + 60y \leq 480$ solves to $y \leq 8 - 1/2\,x$
$60y \leq -30x + 480$

$$y \leq -\frac{1}{2}x + 8$$

Since the total daily profit must be at least $300,

$60x + 75y \geq 300$ is the second inequality.
$60x + 75y \geq 300$ solves to $y \geq 4 - 4/5\,x$
$75y + 60x \geq 300$
$75y \geq -60x + 300$

$$y \geq -\frac{4}{5}x + 4$$

Graph these 2 inequalities:

$y \leq 8 - 1/2\,x$
$y \geq 4 - 4/5\,x$

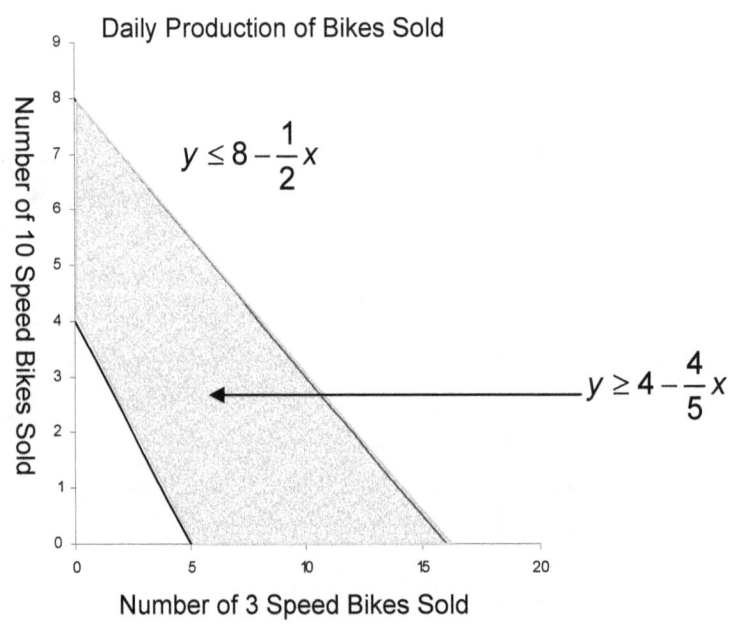

Realize that $x \geq 0$ and $y \geq 0$, since the number of bikes assembled can not be a negative number. Graph these as additional constraints on the problem. The number of bikes assembled must always be an integer value, so points within the shaded area of the graph must have integer values. The maximum profit will occur at or near a corner of the shaded portion of this graph. Those points occur at (0,4), (0,8), (16,0), or (5,0).

Since profits are \$60/3-speed or \$75/10-speed, the profit would be :

$(0,4)$ $60(0) + 75(4) = 300$
$(0,8)$ $60(0) + 75(8) = 600$
$(16,0)$ $60(16) + 75(0) = 960$ ← Maximum profit
$(5,0)$ $60(5) + 75(0) = 300$

The maximum profit would occur if 16 3-speed bikes are made daily.

To solve **a quadratic equation** (with x^2), rewrite the equation into the form:

$$ax^2 + bx + c = 0 \text{ or } y = ax^2 + bx + c$$

where a, b, and c are real numbers. Then substitute the values of a, b, and c into the quadratic formula:

$$x = \frac{-b \pm \sqrt{b^2 - 4ac}}{2a}$$

Simplify the result to find the answers. (Remember, there could be 2 real answers, one real answer, or 2 complex answers that include "i").

To solve a quadratic inequality (with x^2), solve for y. The axis of symmetry is located at $x = {}^-b/2a$. Find coordinates of points to each side of the axis of symmetry. Graph the parabola as a dotted line if the inequality sign is either $<$ or $>$. Graph the parabola as a solid line if the inequality sign is either $\leq$ or $\geq$. Shade above the parabola if the sign is $\geq$ or $>$. Shade below the parabola if the sign is $\leq$ or $<$.

Example: Solve: $8x^2 - 10x - 3 = 0$

In this equation $a = 8$, $b = {}^-10$, and $c = {}^-3$.

Substituting these into the quadratic equation, it becomes:

$$x = \frac{-({}^-10) \pm \sqrt{({}^-10)^2 - 4(8)({}^-3)}}{2(8)} = \frac{10 \pm \sqrt{100 + 96}}{16}$$

$$x = \frac{10 \pm \sqrt{196}}{16} = \frac{10 \pm 14}{16} = 24/16 = 3/2 \text{ or } {}^-4/16 = {}^-1/4$$

Check:

$x = -\dfrac{1}{4}$

$\dfrac{1}{2} + \dfrac{10}{4} - 3 = 0$ Both Check

$3 - 3 = 0$

Example: Solve and graph: $y > x^2 + 4x - 5$.

The axis of symmetry is located at $x = {}^-b/2a$. Substituting 4 for b, and 1 for a, this formula becomes:

$$x = {}^-(4)/2(1) = {}^-4/2 = {}^-2$$

Find coordinates of points to each side of $x = {}^-2$.

x	y
⁻5	0
⁻4	⁻5
⁻3	⁻8
⁻2	⁻9
⁻1	⁻8
0	⁻5
1	0

Graph these points to form a parabola. Draw it as a dotted line. Since a greater than sign is used, shade above and inside the parabola.

MATHEMATICS

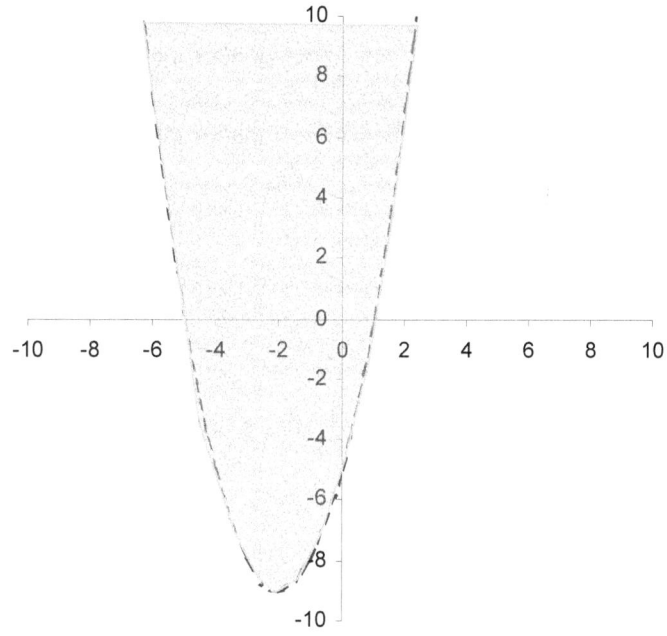

0002   MEASUREMENT

The strategy for solving problems of this nature should be to identify the given shapes and choose the correct formulas. Subtract the smaller cut out shape from the larger shape.

Sample problems:

1. Find the area of one side of the metal in the circular flat washer shown below:

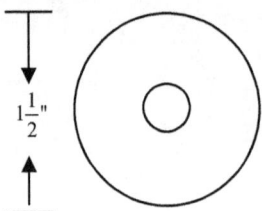

1. the shapes are both circles.

2. use the formula $A = \pi r^2$ for both.

   (Inside diameter is 3/8")

Area of larger circle	Area of smaller circle
$A = \pi r^2$	$A = \pi r^2$
$A = \pi(.75^2)$	$A = \pi(.1875^2)$
$A = 1.76625$ in^2	$A = .1104466$ in^2

Area of metal washer = larger area - smaller area

$= 1.76625$ in$^2 - .1104466$ in^2

$= 1.6558034$ in^2

2. You have decided to fertilize your lawn. The shapes and dimensions of your lot, house, pool and garden are given in the diagram below. The shaded area will not be fertilized. If each bag of fertilizer costs $7.95 and covers 4,500 square feet, find the total number of bags needed and the total cost of the fertilizer.

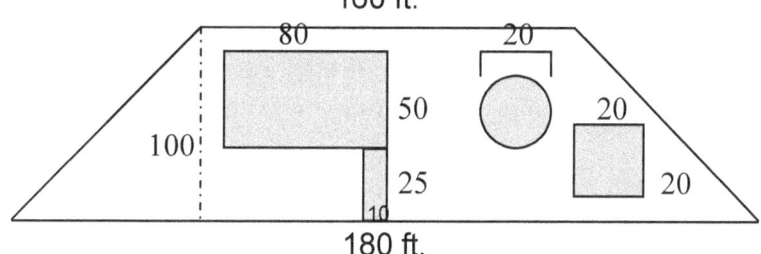

Area of Lot
$A = \frac{1}{2} h(b_1 + b_2)$
$A = \frac{1}{2}(100)(180 + 160)$
$A = 17,000$ sq ft

Area of House
$A = LW$
$A = (80)(50)$
$A = 4,000$ sq ft

Area of Driveway
$A = LW$
$A = (10)(25)$
$A = 250$ sq ft

Area of Pool
$A = \pi r^2$
$A = \pi(10)^2$
$A = 314.159$ sq. ft.

Area of Garden
$A = s^2$
$A = (20)^2$
$A = 400$ sq. ft.

Total area to fertilize = Lot area - (House + Driveway + Pool + Garden)
$= 17,000 - (4,000 + 250 + 314.159 + 400)$
$= 12,035.841$ sq ft

Number of bags needed = Total area to fertilize / 4,500 sq.ft. bag
$= 12,035.841 / 4,500$
$= 2.67$ bags

Since we cannot purchase 2.67 bags we must purchase 3 full bags.

Total cost = Number of bags * $7.95
$= 3 * \$7.95$
$= \$23.85$

Examining the change in area or volume of a given figure requires first to find the existing area given the original dimensions and then finding the new area given the increased dimensions.

Sample problem:

Given the rectangle below determine the change in area if the length is increased by 5 and the width is increased by 7.

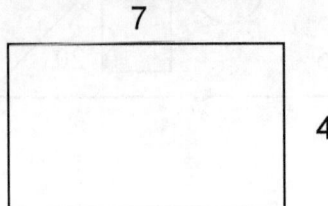

Draw and label a sketch of the new rectangle.

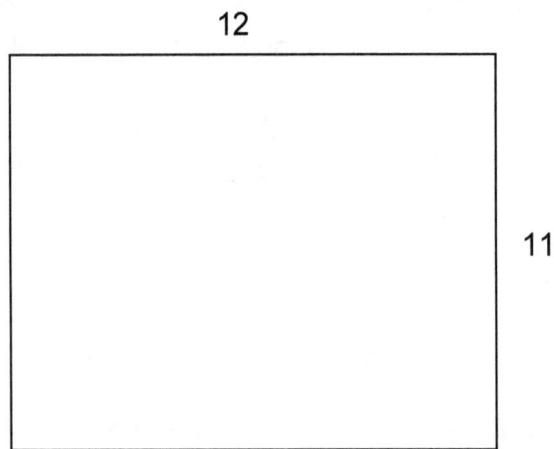

Find the areas.

Area of original = LW   Area of enlarged shape = LW
    = (7)(4)   = (12)(11)
    = 28 units^2 = 132 units^2

The change in area is 132 – 28 = 104 units^2.

Cut the compound shape into smaller, more familiar shapes and then compute the total area by adding the areas of the smaller parts.

Sample problem:

Find the area of the given shape.

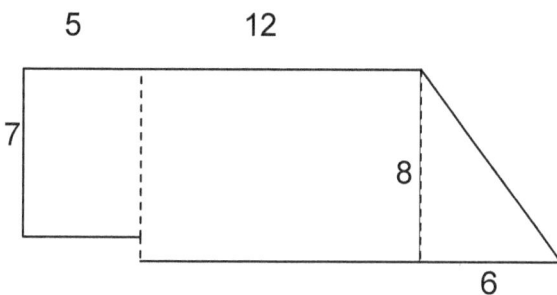

1. Using a dotted line we have cut the shape into smaller parts that are familiar.

2. Use the appropriate formula for each shape and find the sum of all areas.

Area 1 = LW   Area 2 = LW   Area 3 = ½bh
   = (5)(7)   = (12)(8)   = ½(6)(8)
   = 35 units^2 = 96 units^2 = 24 units^2

Total area = Area 1 + Area 2 + Area 3
           = 35 + 96 + 24
           = 155 units^2

It is necessary to be familiar with the metric and customary system in order to estimate measurements.

Some common equivalents include:

ITEM	APPROXIMATELY EQUAL TO	
	METRIC	IMPERIAL
large paper clip	1 gram	1 ounce
1 quart	1 liter	
average sized man	75 kilograms	170 pounds
1 yard	1 meter	
math textbook	1 kilogram	2 pounds
1 mile	1 kilometer	
1 foot	30 centimeters	
thickness of a dime	1 millimeter	0.1 inches

Estimate the measurement of the following items:

The length of an adult cow = _____ meters
The thickness of a compact disc = _____ millimeters
Your height = _____ meters
length of your nose = _____ centimeters
weight of your math textbook = _____ kilograms
weight of an automobile = _____ kilograms
weight of an aspirin = _____ grams

Given a set of objects and their measurements, the use of rounding procedures is helpful when attempting to round to the nearest given unit.

When rounding to a given place value, it is necessary to look at the number in the next smaller place. If this number is 5 or more, the number in the place we are rounding to is increased by one and all numbers to the right are changed to zero. If the number is less than 5, the number in the place we are rounding to stays the same and all numbers to the right are changed to zero.

One method of rounding measurements can require an additional step. First, the measurement must be converted to a decimal number. Then the rules for rounding applied.

Sample problem:

1. Round the measurements to the given units.

MEASUREMENT	ROUND TO NEAREST	ANSWER
1 foot 7 inches	foot	2 ft
5 pound 6 ounces	pound	5 pounds
5 9/16 inches	inch	6 inches

Solution:

Convert each measurement to a decimal number. Then apply the rules for rounding.

1 foot 7 inches = $1\frac{7}{12}$ ft = 1.58333 ft, round up to 2 ft

5 pounds 6 ounces = $5\frac{6}{16}$ pounds = 5.375 pound, round to 5 pounds

$5\frac{9}{16}$ inches = 5.5625 inches, round up to 6 inches

There are many methods for converting measurements within a system. One method is to multiply the given measurement by a conversion factor. This conversion factor is the ratio of:

$$\frac{\text{new units}}{\text{old units}} \quad \text{OR} \quad \frac{\text{what you want}}{\text{what you have}}$$

Sample problems:

1. Convert 3 miles to yards.

$$\frac{3 \text{ miles}}{1} \times \frac{1{,}760 \text{ yards}}{1 \text{ mile}} = \frac{\text{yards}}{}$$

       1. multiply by the conversion factor

$$= 5{,}280 \text{ yards}$$

       2. cancel the miles units
       3. solve

2. Convert 8,750 meters to kilometers.

$$\frac{8{,}750 \text{ meters}}{1} \times \frac{1 \text{ kilometer}}{1000 \text{ meters}} = \frac{\text{km}}{}$$

       1. multiply by the conversion factor

$$= 8.75 \text{ kilometers}$$

       2. cancel the meters units
       3. solve

Most numbers in mathematics are "exact" or "counted". Measurements are "approximate". They usually involve interpolation or figuring out which mark on the ruler is closest. Any measurement you get with a measuring device is approximate. Variations in measurement are called precision and accuracy.

Precision is a measurement of how exactly a measurement is made, without reference to a true or real value. If a measurement is precise it can be made again and again with little variation in the result. The precision of a measuring device is the smallest fractional or decimal division on the instrument. The smaller the unit or fraction of a unit on the measuring device, the more precisely it can measure.

The greatest possible error of measurement is always equal to one-half the smallest fraction of a unit on the measuring device.

Accuracy is a measure of how close the result of measurement comes to the "true" value.

If you are throwing darts, the true value is the bull's eye. If the three darts land on the bull's eye, the dart thrower is both precise (all land near the same spot) and accurate (the darts all land on the "true" value). The greatest measure of error allowed is called the tolerance. The least acceptable limit is called the lower limit and the greatest acceptable limit is called the upper limit. The difference between the upper and lower limits is called the tolerance interval. For example, a specification for an automobile part might be 14.625 ± 0.005 mm. This means that the smallest acceptable length of the part is 14.620 mm and the largest length acceptable is 14.630 mm. The tolerance interval is 0.010 mm. One can see how it would be important for automobile parts to be within a set of limits in terms of length. If the part is too long or too short it will not fit properly and vibrations will occur weakening the part and eventually causing damage to other parts.

0003 GEOMETRY

The equation of a circle with its center at (h,k) and a radius r units is:

$$(x-h)^2 + (y-k)^2 = r^2$$

Sample Problem:

1. Given the equation $x^2 + y^2 = 9$, find the center and the radius of the circle. Then graph the equation.

First, writing the equation in standard circle form gives:

$$(x-0)^2 + (y-0)^2 = 3^2$$

therefore, the center is (0,0) and the radius is 3 units.

Sketch the circle:

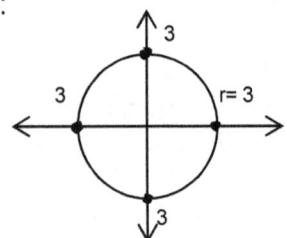

2. Given the equation $x^2 + y^2 - 3x + 8y - 20 = 0$, find the center and the radius. Then graph the circle.

First, write the equation in standard circle form by completing the square for both variables.

$x^2 + y^2 - 3x + 8y - 20 = 0$ 1. Complete the squares.

$(x^2 - 3x + 9/4) + (y^2 + 8y + 16) = 20 + 9/4 + 16$
$(x - 3/2)^2 + (y + 4)^2 = 153/4$

The center is $(3/2, {}^-4)$ and the radius is $\dfrac{\sqrt{153}}{2}$ or $\dfrac{3\sqrt{17}}{2}$.

Graph the circle.

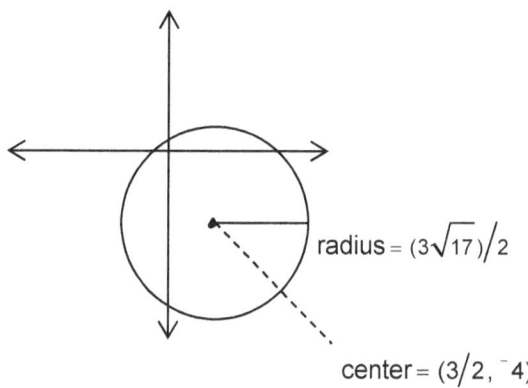

To write the equation given the center and the radius use the **standard form of the equation** of a circle:

$$(x-h)^2 + (y-k)^2 = r^2$$

Sample problems:

Given the center and radius, write the equation of the circle.

1. Center $(^-1, 4)$; radius 11

$(x-h)^2 + (y-k)^2 = r^2$ 1. Write standard equation.

$(x-(^-1))^2 + (y-(4))^2 = 11^2$ 2. Substitute.

$(x+1)^2 + (y-4)^2 = 121$ 3. Simplify.

2. Center $(\sqrt{3}, {^-}1/2)$; radius $= 5\sqrt{2}$

$(x-h)^2 + (y-k)^2 = r^2$ 1. Write standard equation.

$(x-\sqrt{3})^2 + (y-(^-1/2))^2 = (5\sqrt{2})^2$ 2. Substitute.

$(x-\sqrt{3})^2 + (y+1/2)^2 = 50$ 3. Simplify.

In order to accomplish the task of finding the **distance** from a given point to another given line the perpendicular line that intersects the point and line must be drawn and the equation of the other line written. From this information the point of intersection can be found.

This point and the original point are used in the **distance formula** given below:

$$D = \sqrt{(x_2 - x_1)^2 + (y_2 - y_1)^2}$$

Sample Problem:

1. Given the point $(^-4,3)$ and the line $y = 4x + 2$, find the distance from the point to the line.

$y = 4x + 2$	1. Find the slope of the given line by solving for y.
$y = 4x + 2$	2. The slope is 4/1, the perpendicular line will have a slope of $^-1/4$.
$y = (^-1/4)x + b$	3. Use the new slope and the given point to find the equation of the perpendicular line.
$3 = (^-1/4)(^-4) + b$	4. Substitute $(^-4,3)$ into the equation.
$3 = 1 + b$	5. Solve.
$2 = b$	6. Given the value for b, write the equation of the perpendicular line.
$y = (^-1/4)x + 2$	7. Write in standard form.
$x + 4y = 8$	8. Use both equations the point of intersection.
$^-4x + y = 2$	
$x + 4y = 8$	9. Multiply the bottom row by 4.
$^-4x + y = 2$	
$4x + 16y = 32$	10. Solve.
$17y = 34$	
$y = 2$	
$y = 4x + 2$	11. Substitute to find the x value.
$2 = 4x + 2$	12. Solve.
$x = 0$	

(0,2) is the point of intersection. Use this point on the original line and the original point to calculate the distance between them.

$$D = \sqrt{(x_2 - x_1)^2 + (y_2 - y_1)^2}$$ where points are (0,2) and (-4,3).

$$D = \sqrt{(^-4 - 0)^2 + (3 - 2)^2}$$ 1. Substitute.

$$D = \sqrt{(16) + (1)}$$ 2. Simplify.

$$D = \sqrt{17}$$

The **distance between two parallel lines**, such as line AB and line CD as shown below is the line segment RS, the perpendicular between the two parallels.

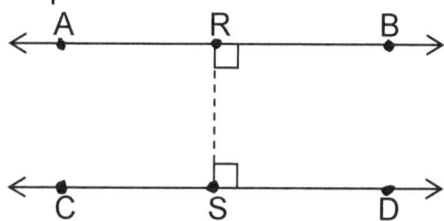

Sample Problem:

Given the geometric figure below, find the distance between the two parallel sides AB and CD.

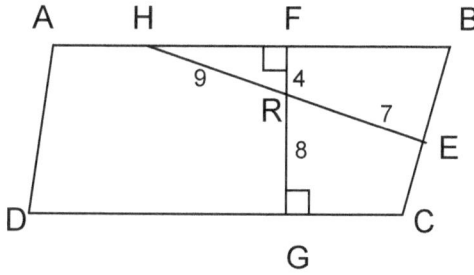

The distance FG is 12 units.

The key to applying the distance formula is to understand the problem before beginning.

$$D = \sqrt{(x_2 - x_1)^2 + (y_2 - y_1)^2}$$

Sample Problem:

1. Find the perimeter of a figure with vertices at (4,5), (⁻4,6) and (⁻5,⁻8).

The figure being described is a triangle. Therefore, the distance for all three sides must be found. Carefully, identify all three sides before beginning.

Side 1 = (4,5) to (⁻4,6)
Side 2 = (⁻4,6) to (⁻5,⁻8)
Side 3 = (⁻5,⁻8) to (4,5)

$$D_1 = \sqrt{(^-4 - 4)^2 + (6 - 5)^2} = \sqrt{65}$$

$$D_2 = \sqrt{((^-5 - (^-4))^2 + (^-8 - 6)^2} = \sqrt{197}$$

$$D_3 = \sqrt{((4 - (^-5))^2 + (5 - (^-8))^2} = \sqrt{250} \text{ or } 5\sqrt{10}$$

$$\text{Perimeter} = \sqrt{65} + \sqrt{197} + 5\sqrt{10}$$

Midpoint Definition:

If a line segment has endpoints of (x_1, y_1) and (x_2, y_2), then the midpoint can be found using:

$$\left(\frac{x_1 + x_2}{2}, \frac{y_1 + y_2}{2} \right)$$

Sample problems:

1. Find the center of a circle with a diameter whose endpoints are $(3,7)$ and $(^-4,^-5)$.

$$\text{Midpoint} = \left(\frac{3+(^-4)}{2}, \frac{7+(^-5)}{2}\right)$$

$$\text{Midpoint} = \left(\frac{^-1}{2}, 1\right)$$

2. Find the midpoint given the two points $\left(5, 8\sqrt{6}\right)$ and $\left(9, ^-4\sqrt{6}\right)$.

$$\text{Midpoint} = \left(\frac{5+9}{2}, \frac{8\sqrt{6}+(^-4\sqrt{6})}{2}\right)$$

$$\text{Midpoint} = \left(7, 2\sqrt{6}\right)$$

Congruent figures have the same size and shape. If one is placed above the other, it will fit exactly. Congruent lines have the same length. Congruent angles have equal measures.
The symbol for congruent is $\cong$.

Polygons (pentagons) *ABCDE* and *VWXYZ* are congruent. They are exactly the same size and shape.

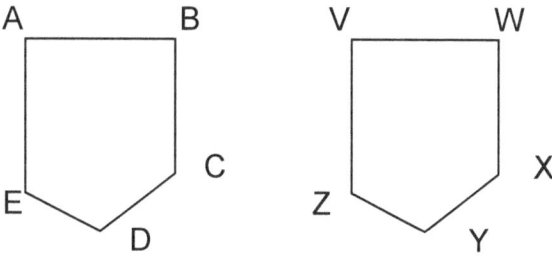

ABCDE $\cong$ *VWXYZ*

Corresponding parts are those congruent angles and congruent sides, that is:

corresponding angles	corresponding sides
$\angle A \leftrightarrow \angle V$	$AB \leftrightarrow VW$
$\angle B \leftrightarrow \angle W$	$BC \leftrightarrow WX$
$\angle C \leftrightarrow \angle X$	$CD \leftrightarrow XY$
$\angle D \leftrightarrow \angle Y$	$DE \leftrightarrow YZ$
$\angle E \leftrightarrow \angle Z$	$AE \leftrightarrow VZ$

Two triangles can be proven congruent by comparing pairs of appropriate congruent corresponding parts.

SSS POSTULATE

If three sides of one triangle are congruent to three sides of another triangle, then the two triangles are congruent.

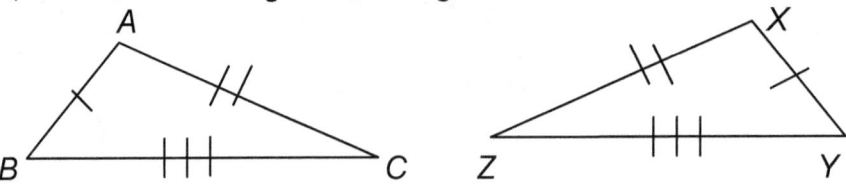

Since $AB \cong XY$, $BC \cong YZ$ and $AC \cong XZ$, then $\triangle ABC \cong \triangle XYZ$.

Example: Given isosceles triangle ABC with D the midpoint of base AC, prove the two triangles formed by AD are congruent.

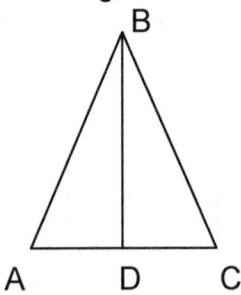

Proof:
1. Isosceles triangle ABC,
   D midpoint of base AC Given
2. $AB \cong BC$ An isosceles $\triangle$ has two congruent sides
3. $AD \cong DC$ Midpoint divides a line into two equal parts
4. $BD \cong BD$ Reflexive
5. $\triangle ABD \cong \triangle BCD$ SSS

SAS POSTULATE

If two sides and the included angle of one triangle are congruent to two sides and the included angle of another triangle, then the two triangles are congruent.

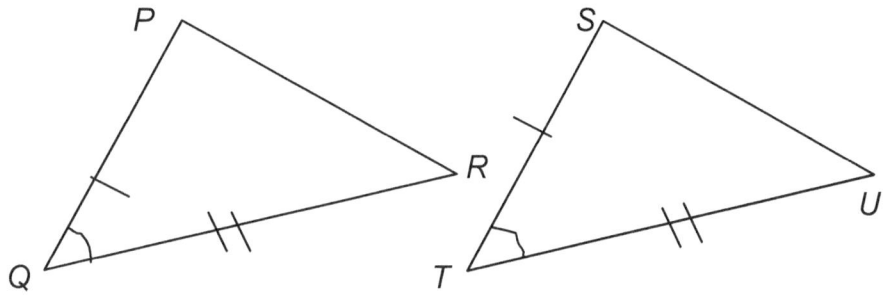

Example:

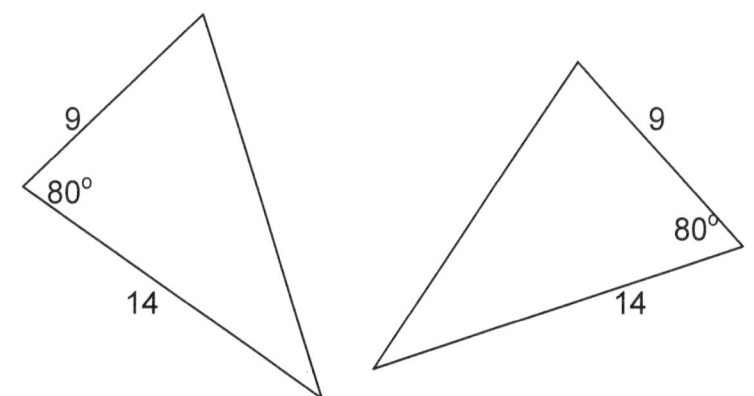

The two triangles are congruent by SAS.

ASA POSTULATE

If two angles and the included side of one triangle are congruent to two angles and the included side of another triangle, the triangles are congruent.

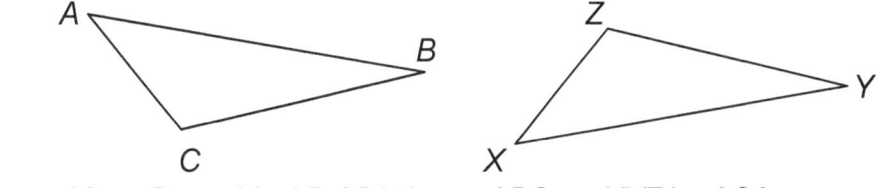

$\angle A \cong \angle X$, $\angle B \cong \angle Y$, $AB \cong XY$ then $\triangle ABC \cong \triangle XYZ$ by ASA

Example 1: Given two right triangles with one leg of each measuring 6 cm and the adjacent angle 37°, prove the triangles are congruent.

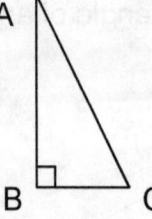
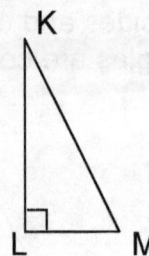

1. Right triangles ABC and KLM   Given
   AB = KL = 6 cm
   ∠A = ∠K = 37°
2. AB ≅ KL   Figures with the
                                        same measure
   ∠A ≅ ∠K   are congruent.
3. ∠B ≅ ∠L   All right angles are
                                        congruent.
4. △ABC ≅ △KLM   ASA

Example 2:
What method would you use to prove the triangles congruent?

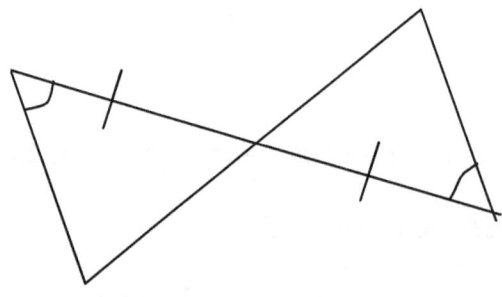

ASA because vertical angles are congruent.

AAS THEOREM

If two angles and a non-included side of one triangle are congruent to the corresponding parts of another triangle, then the triangles are congruent.

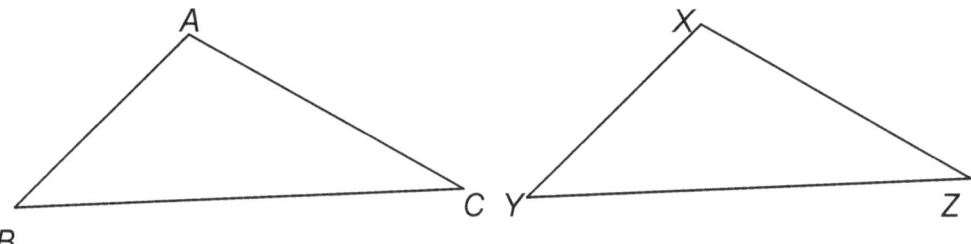

∠B ≅ ∠Y, ∠C ≅ ∠Z, AC ≅ XZ, then △ABC ≅ △XYZ by AAS.

We can derive this theorem because if two angles of the triangles are congruent, then the third angle must also be congruent. Therefore, we can use the ASA postulate.

HL THEOREM

If the hypotenuse and a leg of one right triangle are congruent to the corresponding parts of another right triangle, the triangles are congruent.

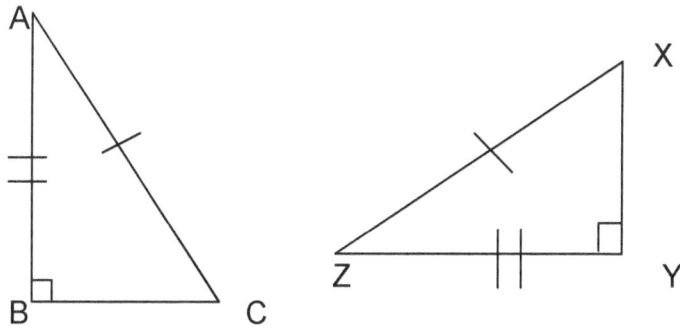

Since ∠B and ∠Y are right angles and AC ≅ XZ (hypotenuse of each triangle), AB ≅ YZ (corresponding leg of each triangle), then △ABC ≅ △XYZ by HL.

Example: What method would you use to prove the triangles congruent?

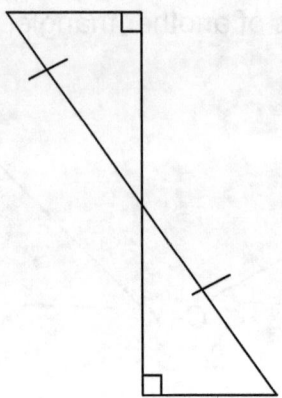

AAS

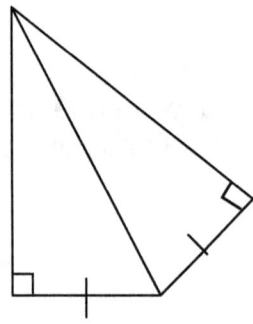

HL

Two figures that have the same shape are similar. Two polygons are similar if corresponding angles are congruent and corresponding sides are in proportion. Corresponding parts of similar polygons are proportional.

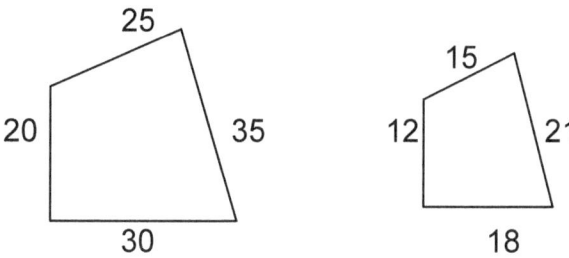

SIMILAR TRIANGLES

AA Similarity Postulate

If two angles of one triangle are congruent to two angles of another triangle, then the triangles are similar.

SAS Similarity Theorem

If an angle of one triangle is congruent to an angle of another triangle and the sides adjacent to those angles are in proportion, then the triangles are similar.

SSS Similarity Theorem

If the sides of two triangles are in proportion, then the triangles are similar.

Example:

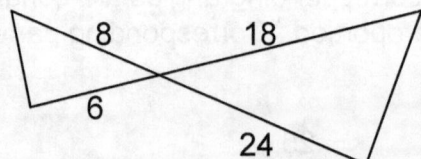

The two triangles are similar since the sides are proportional and vertical angles are congruent.

Example: Given two similar quadrilaterals. Find the lengths of sides *x*, *y*, and *z*.

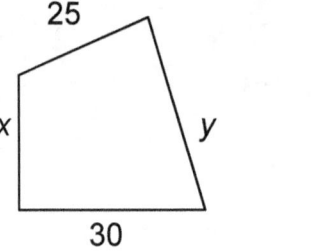 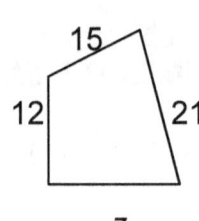

Since corresponding sides are proportional:

= so the scale is

$$\frac{12}{x} = \frac{3}{5} \qquad \frac{21}{y} = \frac{3}{5} \qquad \frac{z}{30} = \frac{3}{5}$$

$$\begin{array}{ccc} 3x = 60 & 3y = 105 & 5z = 90 \\ x = 20 & y = 35 & z = 18 \end{array}$$

A **right triangle** is a triangle with one right angle. The side opposite the right angle is called the **hypotenuse**. The other two sides are the **legs**. An **altitude** is a line drawn from one vertex, perpendicular to the opposite side.

When an altitude is drawn to the hypotenuse of a right triangle, then the two triangles formed are similar to the original triangle and to each other.

Example:

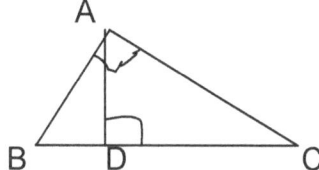

Given right triangle ABC with right angle at A,
altitude AD drawn to hypotenuse BC at D.

$\triangle ABC \sim \triangle ABD \sim \triangle ACD$ The triangles formed are similar to each other and to the original right triangle.

If a, b and c are positive numbers such that $\dfrac{a}{b} = \dfrac{b}{c}$
then b is called the **geometric mean** between a and c.

Example:

Find the geometric mean between 6 and 30.

$$\dfrac{6}{x} = \dfrac{x}{30}$$
$$x^2 = 180$$
$$x = \sqrt{180} = \sqrt{36 \cdot 5} = 6\sqrt{5}$$

The geometric mean is significant when the altitude is drawn to the hypotenuse of a right triangle.
The length of the altitude is the geometric mean between each segment of the hypotenuse,
    and
Each leg is the geometric mean between the hypotenuse and the segment of the hypotenuse that is adjacent to the leg.

Example:

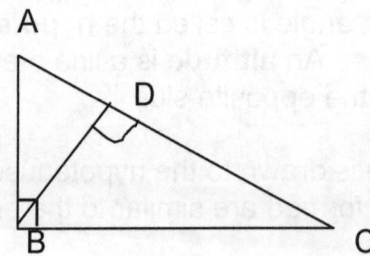

$\triangle ABC$ is a right $\triangle$
BD is the altitude of $\triangle ABC$
AB = 6
AC = 12
Find AD, CD, BD, and BC

$\dfrac{12}{6} = \dfrac{6}{AD}$ $\dfrac{3}{BD} = \dfrac{BD}{9}$ $\dfrac{12}{BC} = \dfrac{BC}{9}$

12(AD) = 36 $(BD)^2 = 27$ $(BC)^2 = 108$

AD = 3

BD = $\sqrt{27}$ = $\sqrt{9 \cdot 3}$ = $3\sqrt{3}$

BC = $\sqrt{108}$ = $\sqrt{36 \cdot 3}$ = $6\sqrt{3}$

CD = 12 - 3 = 9

The **Pythagorean theorem** states that the square of the length of the hypotenuse is equal to the sum of the squares of the lengths of the legs. Symbolically, this is stated as:

$$c^2 = a^2 + b^2$$

Given the right triangle below, find the missing side.

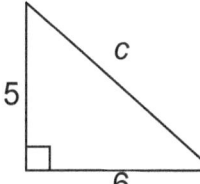

$c^2 = a^2 + b^2$	1. write formula
$c^2 = 5^2 + 6^2$	2. substitute known values
$c^2 = 61$	3. take square root
$c = \sqrt{61}$ or 7.81	4. solve

The **Converse of the Pythagorean Theorem** states that if the square of one side of a triangle is equal to the sum of the squares of the other two sides, then the triangle is a right triangle.

Example:
Given △XYZ, with sides measuring 12, 16 and 20 cm. Is this a right triangle?

$$c^2 = a^2 + b^2$$
$$20^2 \ ? \ 12^2 + 16^2$$
$$400 \ ? \ 144 + 256$$
$$400 = 400$$

Yes, the triangle is a right triangle.

This theorem can be expanded to determine if triangles are obtuse or acute.

If the square of the longest side of a triangle is greater than the sum of the squares of the other two sides, then the triangle is an obtuse triangle.
and
If the square of the longest side of a triangle is less than the sum of the squares of the other two sides, then the triangle is an acute triangle.

Example:
Given △LMN with sides measuring 7, 12, and 14 inches. Is the triangle right, acute, or obtuse?

14^2 ? $7^2 + 12^2$
196 ? 49 + 144
196 > 193

Therefore, the triangle is obtuse.

Given the **special right triangles below**, we can find the lengths of other special right triangles.

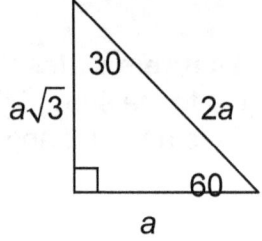 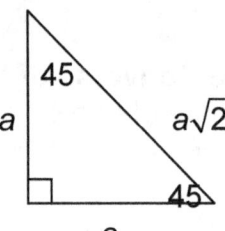

Sample problems:

1. 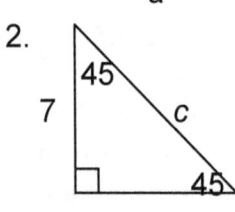   if $8 = a\sqrt{2}$ then $a = 8/\sqrt{2}$ or 5.657

2. 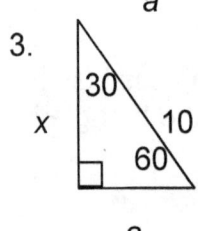   if $7 = a$ then $c = a\sqrt{2} = 7\sqrt{2}$ or 9.899

3. if $2a = 10$ then $a = 5$ and $x = a\sqrt{3} = 5\sqrt{3}$ or 8.66

MATHEMATICS 52

Given right triangle right ABC, the adjacent side and opposite side can be identified for each angle A and B.

Looking at angle A, it can be determined that side *b* is adjacent to angle A and side *a* is opposite angle A.

If we now look at angle B, we see that side *a* is adjacent to angle *b* and side *b* is opposite angle B.

The longest side (opposite the 90 degree angle) is always called the hypotenuse.

The basic trigonometric ratios are listed below:

Sine = opposite/hypotenuse   Cosine = adjacent/hypotenuse   Tangent = opposite/adjacent

Sample problem:

1. Use triangle ABC to find the sin, cos and tan for angle A.

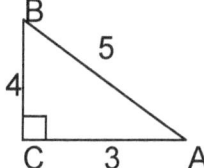

Sin A = 4/5
Cos A = 3/5
Tan A = 4/3

Use the basic trigonometric ratios of sine, cosine and tangent to solve for the missing sides of right triangles when given at least one of the acute angles.

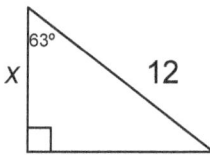

In the triangle ABC, an acute angle of 63 degrees and the length of the hypotenuse (12). The missing side is the one adjacent to the given angle.

The appropriate trigonometric ratio to use would be cosine since we are looking for the adjacent side and we have the length of the hypotenuse.

$Cos x = \dfrac{\text{adjacent}}{\text{hypotenuse}}$ 1. Write formula.

$Cos 63 = \dfrac{x}{12}$ 2. Substitute known values.

$0.454 = \dfrac{x}{12}$

$x = 5.448$ 3. Solve.

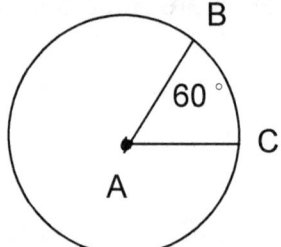

Central angle BAC = 60°
Minor arc BC = 60°
Major arc BC = 360 − 60 = 300°

If you draw **two radii** in a circle, the angle they form with the center as the vertex is a central angle. The piece of the circle "inside" the angle is an arc. Just like a central angle, an arc can have any degree measure from 0 to 360. The measure of an arc is equal to the measure of the central angle which forms the arc. Since a diameter forms a semicircle and the measure of a straight angle like a diameter is 180°, the measure of a semicircle is also 180°.

Given two points on a circle, there are two different arcs which the two points form. Except in the case of semicircles, one of the two arcs will always be greater than 180° and the other will be less than 180°. The arc less than 180° is a minor arc and the arc greater than 180° is a major arc.

Examples:

1.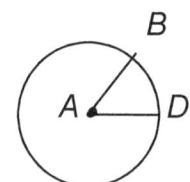

   $m\angle BAD = 45°$
   What is the measure of the major arc BD?

   $\angle BAD$ = minor arc BD

   $45°$ = minor arc BD

   $360 - 45$ = major arc BD

   $315°$ = major arc BD

   The measure of the central angle is the same as the measure of the arc it forms.
   A major and minor arc always add to $360°$.

2.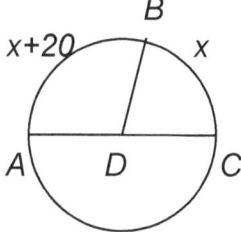

   $\overline{AC}$ is a diameter of circle D. What is the measure of $\angle BDC$?

   $m\angle ADB + m\angle BDC = 180°$
   $x + 20 + x = 180$
   $2x + 20 = 180$
   $2x = 160$
   $x = 80$
   minor arc $BC = 80°$
   $m\angle BDC = 80°$

   A diameter forms a semicircle which has a measure of $180°$.

   A central angle has the same measure as the arc it forms.

MATHEMATICS

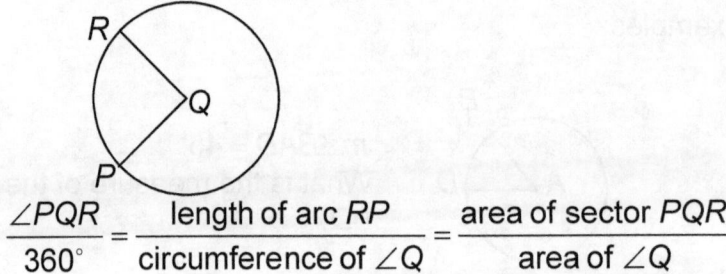

$$\frac{\angle PQR}{360°} = \frac{\text{length of arc } RP}{\text{circumference of } \angle Q} = \frac{\text{area of sector } PQR}{\text{area of } \angle Q}$$

While an arc has a measure associated to the degree measure of a central angle, it also has a length which is a fraction of the circumference of the circle.

For each central angle and its associated arc, there is a sector of the circle which resembles a pie piece. The area of such a sector is a fraction of the area of the circle.

The fractions used for the area of a sector and length of its associated arc are both equal to the ratio of the central angle to 360°.

Examples:
1.

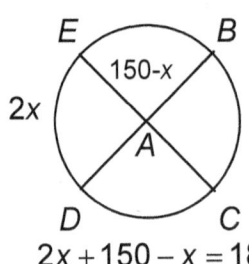

$\odot A$ has a radius of 4 cm. What is the length of arc ED?

$2x + 150 - x = 180$

$x + 150 = 180$

$x = 30$

Arc $ED = 2(30) = 60°$

Arc BE and arc DE make a semicircle.

The ratio 60° to 360° is equal to the ratio of arch length ED to the circumference of $\odot A$.

$$\frac{60}{360} = \frac{\text{arc length } ED}{2\pi 4}$$

$$\frac{1}{6} = \frac{\text{arc length}}{8\pi}$$

Cross multiply and solve for the arc length.

$$\frac{8\pi}{6} = \text{arc length}$$

arc length $ED = \dfrac{4\pi}{3}$ cm

2.

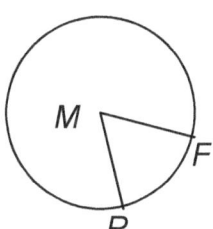

The radius of ⊙M is 3 cm. The length of arc PF is 2π cm. What is the area of sector PMF?

Circumference of ⊙$M = 2\pi(3) = 6\pi$

Area of ⊙$M = \pi(3)^2 = 9\pi$

$$\frac{\text{area of } PMF}{9\pi} = \frac{2\pi}{6\pi}$$

Find the circumference and area of the circle.

The ratio of the sector area to the circle area is the same as the arc length to the circumference.

$$\frac{\text{area of } PMF}{9\pi} = \frac{1}{3}$$

area of $PMF = \dfrac{9\pi}{3}$

area of $PMF = 3\pi$

Solve for the area of the sector.

A tangent line intersects a circle in exactly one point. If a radius is drawn to that point, the radius will be perpendicular to the tangent.

A chord is a segment with endpoints on the circle. If a radius or diameter is perpendicular to a chord, the radius will cut the chord into two equal parts.

If **two chords** in the same circle have the same length, the two chords will have arcs that are the same length, and the two chords will be equidistant from the center of the circle. Distance from the center to a chord is measured by finding the length of a segment from the center perpendicular to the chord.

Examples:

1.

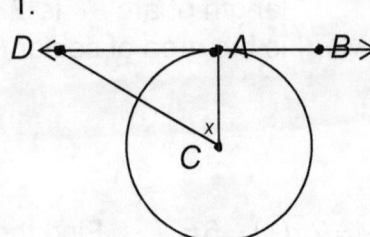

$\overrightarrow{DB}$ is tangent to $\odot C$ at A.
$m \angle ADC = 40°$. Find x.

$\overline{AC} \perp \overrightarrow{DB}$ A radius is $\perp$ to a tangent at the point of tangency.

$m \angle DAC = 90°$ Two segments that are $\perp$ form a $90°$ angle.

$40 + 90 + x = 180$ The sum of the angles of a triangle is $180°$.

$x = 50°$ Solve for x.

2.

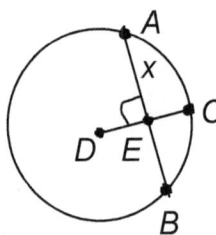

$\overline{CD}$ is a radius and $\overline{CD} \perp$ chord $\overline{AB}$.
$\overline{AB} = 10$. Find x.

$x = \dfrac{1}{2}(10)$

$x = 5$ If a radius is $\perp$ to a chord, the radius bisects the chord.

Angles with their vertices on the circle:

An inscribed angle is an angle whose vertex is on the circle. Such an angle could be formed by two chords, two diameters, two secants, or a secant and a tangent. An inscribed angle has one arc of the circle in its interior. The measure of the inscribed angle is one-half the measure of this intercepted arc. If two inscribed angles intercept the same arc, the two angles are congruent (i.e. their measures are equal). If an inscribed angle intercepts an entire semicircle, the angle is a right angle.

Angles with their vertices in a circle's interior:

When two chords intersect inside a circle, two sets of vertical angles are formed. Each set of vertical angles intercepts two arcs which are across from each other. The measure of an angle formed by two chords in a circle is equal to one-half the sum of the angle intercepted by the angle and the arc intercepted by its vertical angle.

Angles with their vertices in a circle's exterior:

If an angle has its vertex outside of the circle and each side of the circle intersects the circle, then the angle contains two different arcs. The measure of the angle is equal to one-half the difference of the two arcs.

Examples:

1.

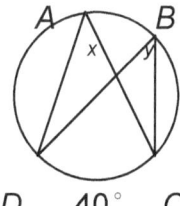

Find x and y.
arc $DC = 40°$

$m\angle DAC = \dfrac{1}{2}(40) = 20°$

$\angle DAC$ and $\angle DBC$ are both inscribed angles, so each one has a measure equal to one-half the measure of arc DC.

$m\angle DBC = \dfrac{1}{2}(40) = 20°$

$x = 20°$ and $y = 20°$

Intersecting chords:

If two chords intersect inside a circle, each chord is divided into two smaller segments. The product of the lengths of the two segments formed from one chord equals the product of the lengths of the two segments formed from the other chord.

Intersecting tangent segments:

If two tangent segments intersect outside of a circle, the two segments have the same length.

Intersecting secant segments:

If two secant segments intersect outside a circle, a portion of each segment will lie inside the circle and a portion (called the exterior segment) will lie outside the circle. The product of the length of one secant segment and the length of its exterior segment equals the product of the length of the other secant segment and the length of its exterior segment.

Tangent segments intersecting secant segments:

If a tangent segment and a secant segment intersect outside a circle, the square of the length of the tangent segment equals the product of the length of the secant segment and its exterior segment.
Examples:

1.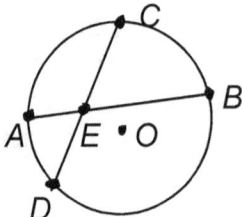

$\overline{AB}$ and $\overline{CD}$ are chords.
$CE=10$, $ED=x$, $AE=5$, $EB=4$

$(AE)(EB) = (CE)(ED)$ Since the chords intersect in the circle, the products of the segment pieces are equal.

$5(4) = 10x$
$20 = 10x$
$x = 2$ Solve for x.

2.

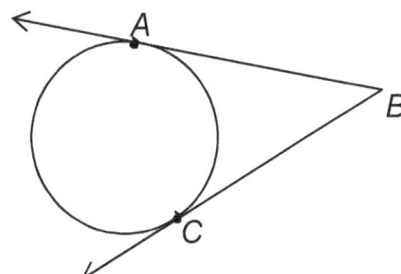

$\overline{AB}$ and $\overline{CD}$ are chords.
$\overline{AB} = x^2 + x - 2$
$\overline{BC} = x^2 - 3x + 5$
Find the length of
$\overline{AB}$ and $\overline{BC}$.

$\overline{AB} = x^2 + x - 2$
$\overline{BC} = x^2 - 3x + 5$ Given

$\overline{AB} = \overline{BC}$ Intersecting tangents are equal.

$x^2 + x - 2 = x^2 - 3x + 5$ Set the expression equal to each other and solve.

$4x = 7$
$x = 1.75$ Substitute and solve.

$(1.75)^2 + 1.75 - 2 = \overline{AB}$
$\overline{AB} = \overline{BC} = 2.81$

Two triangles are congruent if each of the three angles and three sides of one triangle match up in a one-to-one fashion with congruent angles and sides of the second triangle. In order to see how the sides and angles match up, it is sometimes necessary to imagine rotating or reflecting one of the triangles so the two figures are oriented in the same position.

There are shortcuts to the above procedure for proving two triangles congruent.

Side-Side-Side (SSS) Congruence--If the three sides of one triangle match up in a one-to-one congruent fashion with the three sides of the other triangle, then the two triangles are congruent. With SSS it is not necessary to even compare the angles; they will automatically be congruent.

Angle-Side-Angle (ASA) Congruence--If two angles of one triangle match up in a one-to-one congruent fashion with two angles in the other triangle and if the sides between the two angles are also congruent, then the two triangles are congruent. With ASA the sides that are used for congruence must be located between the two angles used in the first part of the proof.

Side-Angle-Side (SAS) Congruence--If two sides of one triangle match up in a one-to-one congruent fashion with two sides in the other triangle and if the angles between the two sides are also congruent, then the two triangles are congruent. With SAS the angles that are used for congruence must be located between the two sides used in the first part of the proof.

In addition to SSS, ASA, and SAS, **Angle-Angle-Side (AAS)** is also a congruence shortcut.

AAS states that if two angles of one triangle match up in a one-to-one congruent fashion with two angles in the other triangle and if two sides that are not between the aforementioned sets of angles are also congruent, then the triangles are congruent. ASA and AAS are very similar; the only difference is where the congruent sides are located. If the sides are between the congruent sets of angles, use ASA. If the sides are not located between the congruent sets of angles, use AAS.

Hypotenuse-Leg (HL) is a congruence shortcut which can only be used with right triangles. If the hypotenuse and leg of one right triangle are congruent to the hypotenuse and leg of the other right triangle, then the two triangles are congruent.

Two triangles are overlapping if a portion of the interior region of one triangle is shared in common with all or a part of the interior region of the second triangle.

The most effective method for proving two overlapping triangles congruent is to draw the two triangles separated. Separate the two triangles and label all of the vertices using the labels from the original overlapping figures. Once the separation is complete, apply one of the congruence shortcuts: SSS, ASA, SAS, AAS, or HL.

A parallelogram is a quadrilateral (four-sided figure) in which opposite sides are parallel. There are three shortcuts for proving that a quadrilateral is a parallelogram without directly showing that the opposite sides are parallel.

If the diagonals of a quadrilateral bisect each other, then the quadrilateral is also a parallelogram. Note that this shortcut only requires the diagonals to bisect each other; the diagonals do not need to be congruent.

If both pairs of opposite sides are congruent, then the quadrilateral is a parallelogram.

If both pairs of opposite angles are congruent, then the quadrilateral is a parallelogram.

If one pair of opposite sides are both parallel and congruent, then the quadrilateral is a parallelogram.

The following table illustrates the properties of each quadrilateral.

	Parallel Opposite Sides	Bisecting Diagonals	Equal Opposite Sides	Equal Opposite Angles	Equal Diagonals	All Sides Equal	All Angles Equal	Perpendicular Diagonals
Parallelogram	X	X	X	X				
Rectangle	X	X	X	X	X		X	
Rhombus	X	X	X	X		X		X
Square	X	X	X	X	X	X	X	X

A trapezoid is a quadrilateral with exactly one pair of parallel sides. A trapezoid is different from a parallelogram because a parallelogram has two pairs of parallel sides.

The two parallel sides of a trapezoid are called the bases, and the two non-parallel sides are called the legs. If the two legs are the same length, then the trapezoid is called isosceles.

The segment connecting the two midpoints of the legs is called the median. The median has the following two properties.

The median is parallel to the two bases.

The length of the median is equal to one-half the sum of the length of the two bases.

The segment joining the midpoints of two sides of a triangle is called a **median**. All triangles have three medians. Each median has the following two properties.

A median is parallel to the third side of the triangle.
The length of a median is one-half the length of the third side of the triangle.

Every **angle** has exactly one ray which bisects the angle. If a point on such a bisector is located, then the point is equidistant from the two sides of the angle. Distance from a point to a side is measured along a segment which is perpendicular to the angle's side. The converse is also true. If a point is equidistant from the sides of an angle, then the point is on the bisector of the angle.

Every **segment** has exactly one line which is both perpendicular to and bisects the segment. If a point on such a perpendicular bisector is located, then the point is equidistant to the endpoints of the segment. The converse is also true. If a point is equidistant from the endpoints of a segment, then that point is on the perpendicular bisector of the segment.

A median is a segment that connects a vertex to the midpoint of the side opposite from that vertex. Every triangle has exactly three medians.

An altitude is a segment which extends from one vertex and is perpendicular to the side opposite that vertex. In some cases, the side opposite from the vertex used will need to be extended in order for the altitude to form a perpendicular to the opposite side. The length of the altitude is used when referring to the height of the triangle.

If the three segments which bisect the three angles of a triangle are drawn, the segments will all intersect in a single point. This point is equidistant from all three sides of the triangle. Recall that the distance from a point to a side is measured along the perpendicular from the point to the side.

If two planes are parallel and a third plane intersects the first two, then the three planes will intersect in two lines which are also parallel.

Given a line and a point which is not on the line but is in the same plane, then there is exactly one line through the point which is parallel to the given line and exactly one line through the point which is perpendicular to the given line.

If three or more segments intersect in a single point, the point is called a **point of concurrency**.

The following sets of special segments all intersect in points of concurrency.
1. Angle Bisectors
2. Medians
3. Altitudes
4. Perpendicular Bisectors

The points of concurrency can lie inside the triangle, outside the triangle, or on one of the sides of the triangle. The following table summarizes this information.

Possible Location(s) of the
Points of Concurrency

	Inside the Triangle	Outside the Triangle	On the Triangle
Angle Bisectors	x		
Medians	x		
Altitudes	x	x	x
Perpendicular Bisectors	x	x	x

A circle is inscribed in a triangle if the three sides of the triangle are each tangent to the circle. The center of an inscribed circle is called the incenter of the triangle. To find the incenter, draw the three angle bisectors of the triangle. The point of concurrency of the angle bisectors is the incenter or center of the inscribed circle. Each triangle has only one inscribed circle.

A circle is circumscribed about a triangle if the three vertices of the triangle are all located on the circle. The center of a circumscribed circle is called the circumcenter of the triangle. To find the circumcenter, draw the three perpendicular bisectors of the sides of the triangle. The point of concurrency of the perpendicular bisectors is the circumcenter or the center of the circumscribing circle. Each triangle has only one circumscribing circle.

A median is a segment which connects a vertex to the midpoint of the side opposite that vertex. Every triangle has three medians. The point of concurrency of the three medians is called the **centroid**.

The centroid divides each median into two segments whose lengths are always in the ratio of 1:2. The distance from the vertex to the centroid is always twice the distance from the centroid to the midpoint of the side opposite the vertex.

If two circles have radii which are in a ratio of $a:b$, then the following ratios are also true for the circles.

The diameters are also in the ratio of $a:b$.
The circumferences are also in the ratio $a:b$.
The areas are in the ratio $a^2:b^2$, or the ratio of the areas is the square of the ratios of the radii.

In order to determine if a figure is convex and then determine if it is regular, it is necessary to apply the definition of convex first.

Convex polygons: polygons in which no line containing the side of the polygon contains a point on the interior of the polygon.

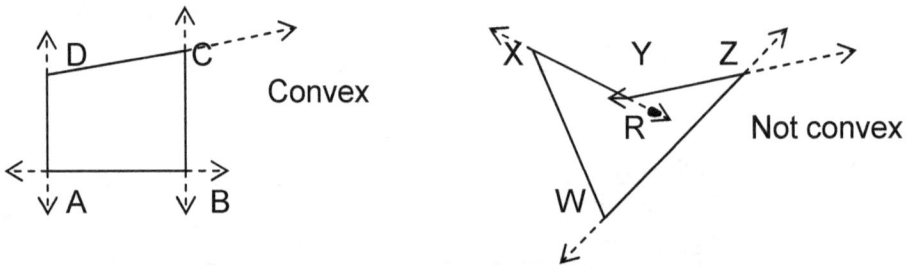

Regular polygons: convex polygons in which all sides are congruent and all angles are congruent (in other words, a regular polygon must be both equilateral and equiangular).

Use the formulas to find the volume and surface area.

FIGURE	VOLUME	TOTAL SURFACE AREA
Right Cylinder	$\pi r^2 h$	$2\pi rh + 2\pi r^2$
Right Cone	$\dfrac{\pi r^2 h}{3}$	$\pi r\sqrt{r^2 + h^2} + \pi r^2$
Sphere	$\dfrac{4}{3}\pi r^3$	$4\pi r^2$
Rectangular Solid	LWH	$2LW + 2WH + 2LH$

Note: $\sqrt{r^2 + h^2}$ is equal to the slant height of the cone.

Sample problem:

1. Given the figure below, find the volume and surface area.

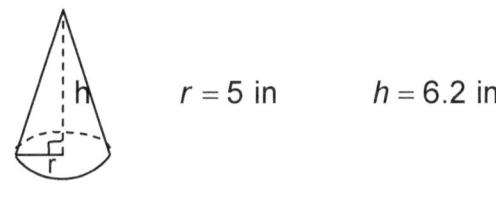

$r = 5$ in $h = 6.2$ in

Volume $= \dfrac{\pi r^2 h}{3}$ First write the formula.

$\dfrac{1}{3}\pi(5^2)(6.2)$ Then substitute.

162.3 cubic inches   Finally solve the problem.

Surface area $= \pi r\sqrt{r^2 + h^2} + \pi r^2$ First write the formula.

$\pi 5\sqrt{5^2 + 6.2^2} + \pi 5^2$ Then substitute.

203.6 square inches   Compute.

Note: volume is always given in cubic units and area is always given in square units.

FIGURE	AREA FORMULA	PERIMETER FORMULA
Rectangle	LW	$2(L+W)$
Triangle	$\frac{1}{2}bh$	$a+b+c$
Parallelogram	bh	sum of lengths of sides
Trapezoid	$\frac{1}{2}h(a+b)$	sum of lengths of sides

Sample problems:
1. Find the area and perimeter of a rectangle if its length is 12 inches and its diagonal is 15 inches.

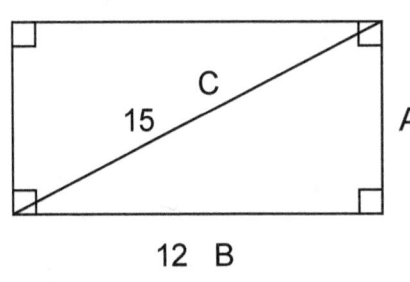

1. Draw and label sketch.

2. Since the height is still needed use Pythagorean formula to find missing leg of the triangle.

$$A^2 + B^2 = C^2$$
$$A^2 + 12^2 = 15^2$$
$$A^2 = 15^2 - 12^2$$
$$A^2 = 81$$
$$A = 9$$

Now use this information to find the area and perimeter.

$A = LW$	$P = 2(L+W)$	1. write formula
$A = (12)(9)$	$P = 2(12+9)$	2. substitute
$A = 108 \text{ in}^2$	$P = 42$ inches	3. solve

Given a circular figure the formulas are as follows:

$A = \pi r^2$ $\qquad$ $C = \pi d$ or $2\pi r$

MATHEMATICS 68

Sample problem:

1. If the area of a circle is 50 cm^2, find the circumference.

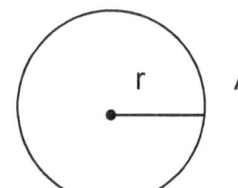
$A = 50$ cm^2

1. Draw sketch.
2. Determine what is still needed.

Use the area formula to find the radius.

$A = \pi r^2$	1. write formula
$50 = \pi r^2$	2. substitute
$\dfrac{50}{\pi} = r^2$	3. divide by π
$15.915 = r^2$	4. substitute
$\sqrt{15.915} = \sqrt{r^2}$	5. take square root of both sides
$3.989 \approx r$	6. compute

Use the approximate answer (due to rounding) to find the circumference.

$C = 2\pi r$	1. write formula
$C = 2\pi (3.989)$	2. substitute
$C \approx 25.064$	3. compute

When **using formulas** to find each of the required items it is helpful to remember to always use the same strategies for problem solving. First, draw and label a sketch if needed. Second, write the formula down and then substitute in the known values. This will assist in identifying what is still needed (the unknown). Finally, solve the resulting equation.

Being consistent in the strategic approach to problem solving is paramount to teaching the concept as well as solving it.

Use appropriate problem solving strategies to find the solution.

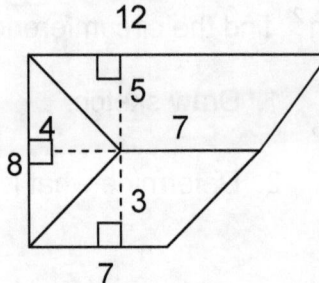

1. Find the area of the given figure.
2. Cut the figure into familiar shapes.
3. Identify what type figures are given and write the appropriate formulas.

Area of figure 1　　　Area of figure 2　　　Area of figure 3
(triangle)　　　　　　(parallelogram)　　　　(trapezoid)

$A = \dfrac{1}{2}bh$　　　　　　$A = bh$　　　　　　　$A = \dfrac{1}{2}h(a+b)$

$A = \dfrac{1}{2}(8)(4)$　　　　$A = (7)(3)$　　　　　$A = \dfrac{1}{2}(5)(12+7)$

$A = 16$ sq. ft　　　　$A = 21$ sq. ft　　　　$A = 47.5$ sq. ft

Now find the total area by adding the area of all figures.

Total area $= 16 + 21 + 47.5$
Total area $= 84.5$ square ft

Given the figure below, find the area by dividing the polygon into smaller shapes.

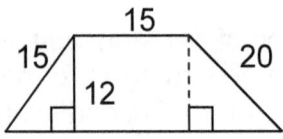

1. divide the figure into two triangles and a rectangle.
2. find the missing lengths.
3. find the area of each part.
4. find the sum of all areas.

Find base of both right triangles using Pythagorean Formula:

$a^2 + b^2 = c^2$　　　　　　$a^2 + b^2 = c^2$
$a^2 + 12^2 = 15^2$　　　　$a^2 + 12^2 = 20^2$
$a^2 = 225 - 144$　　　　　$a^2 = 400 - 144$
$a^2 = 81$　　　　　　　　$a^2 = 256$
$a = 9$　　　　　　　　　　$a = 16$

Area of triangle 1	Area of triangle 2	Area of rectangle
$A = \frac{1}{2}bh$	$A = \frac{1}{2}bh$	$A = LW$
$A = \frac{1}{2}(9)(12)$	$A = \frac{1}{2}(16)(12)$	$A = (15)(12)$
$A = 54$ sq. units	$A = 96$ sq. units	$A = 180$ sq. units

Find the sum of all three figures.

$$54 + 96 + 180 = 330 \text{ square units}$$

Polygons are similar if and only if there is a one-to-one correspondence between their vertices such that the corresponding angles are congruent and the lengths of corresponding sides are proportional.

Given the rectangles below, compare the area and perimeter.

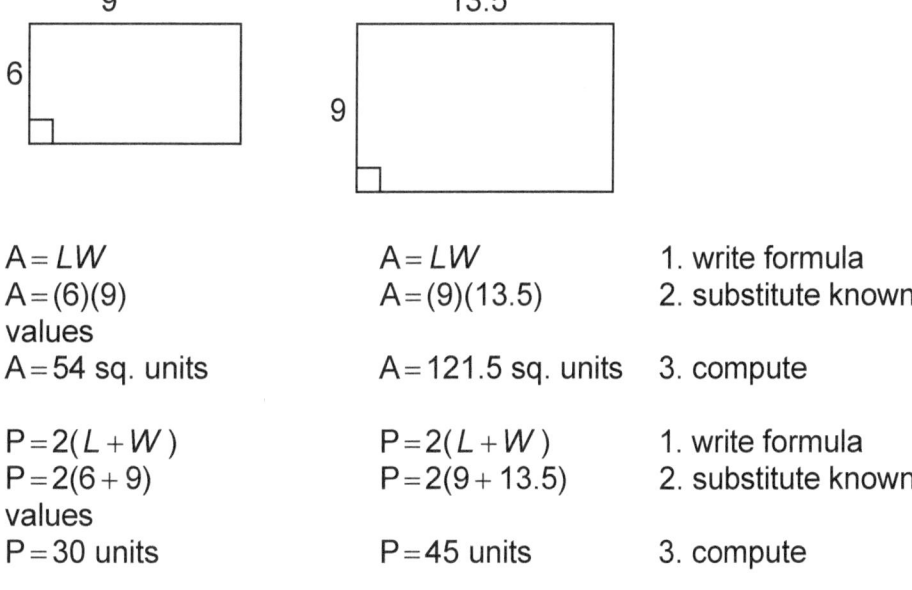

$A = LW$	$A = LW$	1. write formula
$A = (6)(9)$	$A = (9)(13.5)$	2. substitute known values
$A = 54$ sq. units	$A = 121.5$ sq. units	3. compute
$P = 2(L + W)$	$P = 2(L + W)$	1. write formula
$P = 2(6 + 9)$	$P = 2(9 + 13.5)$	2. substitute known values
$P = 30$ units	$P = 45$ units	3. compute

Notice that the areas relate to each other in the following manner:

Ratio of sides $9/13.5 = 2/3$

Multiply the first area by the square of the reciprocal $(3/2)^2$ to get the second area.
$$54 \times (3/2)^2 = 121.5$$

The perimeters relate to each other in the following manner:

Ratio of sides   9/13.5 = 2/3

Multiply the perimeter of the first by the reciprocal of the ratio to get the perimeter of the second.

$$30 \times 3/2 = 45$$

FIGURE	LATERAL AREA	TOTAL AREA	VOLUME
Right prism	sum of area of lateral faces (rectangles)	lateral area plus 2 times the area of base	area of base times height
regular pyramid	sum of area of lateral faces (triangles)	lateral area plus area of base	1/3 times the area of the base times the height

Find the total area of the given figure:

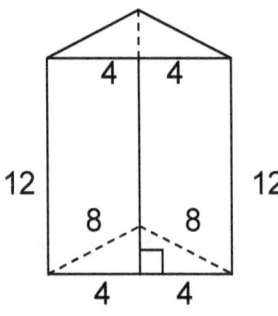

1. Since this is a triangular prism, first find the area of the bases.

2. Find the area of each rectangular lateral face.

3. Add the areas together.

$A = \frac{1}{2}bh$ $A = LW$ 1. write formula

$8^2 = 4^2 + h^2$
$h = 6.928$ 2. find the height of the base triangle

$A = \frac{1}{2}(8)(6.928)$ $A = (8)(12)$

                                            3. substitute known values
$A = 27.713$ sq. units $A = 96$ sq. units   4. compute

Total Area = 2(27.713) + 3(96)
          = 343.426 sq. units

FIGURE	VOLUME	TOTAL SURFACE AREA	LATERAL AREA
Right Cylinder	$\pi r^2 h$	$2\pi rh + 2\pi r^2$	$2\pi rh$
Right Cone	$\dfrac{\pi r^2 h}{3}$	$\pi r\sqrt{r^2 + h^2} + \pi r^2$	$\pi r\sqrt{r^2 + h^2}$

Note: $\sqrt{r^2 + h^2}$ is equal to the slant height of the cone.

Sample problem:

1. A water company is trying to decide whether to use traditional cylindrical paper cups or to offer conical paper cups since both cost the same. The traditional cups are 8 cm wide and 14 cm high. The conical cups are 12 cm wide and 19 cm high. The company will use the cup that holds the most water.

    1. Draw and label a sketch of each.

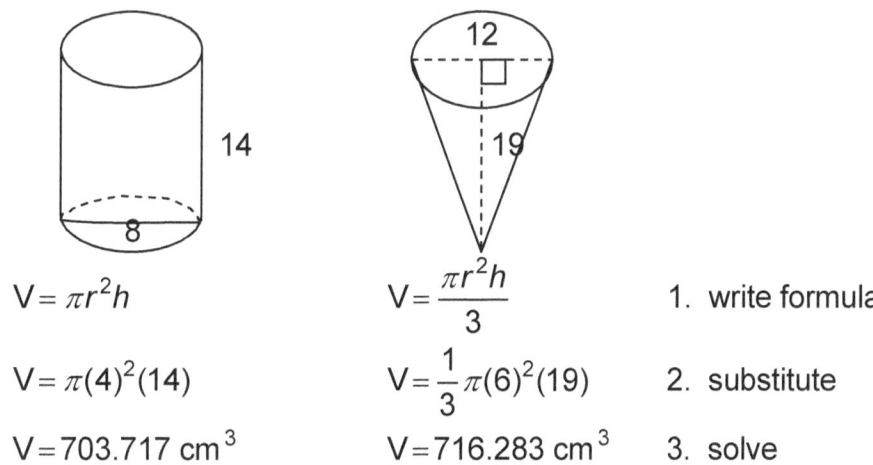

$V = \pi r^2 h$	$V = \dfrac{\pi r^2 h}{3}$	1. write formula
$V = \pi(4)^2(14)$	$V = \dfrac{1}{3}\pi(6)^2(19)$	2. substitute
$V = 703.717 \text{ cm}^3$	$V = 716.283 \text{ cm}^3$	3. solve

The choice should be the conical cup since its volume is more.

FIGURE	VOLUME	TOTAL SURFACE AREA
Sphere	$\frac{4}{3}\pi r^3$	$4\pi r^2$

Sample problem:

1. How much material is needed to make a basketball that has a diameter of 15 inches? How much air is needed to fill the basketball?

Draw and label a sketch:

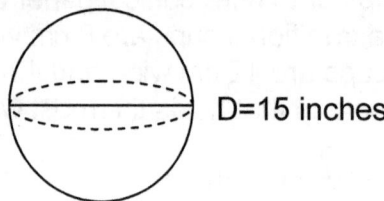

D=15 inches

Total surface area

$TSA = 4\pi r^2$

$= 4\pi(7.5)^2$

$= 706.9 \text{ in}^2$

Volume

$V = \frac{4}{3}\pi r^3$

$= \frac{4}{3}\pi(7.5)^3$

$= 1767.1 \text{ in}^3$

1. write formula
2. substitute
3. solve

0004 TRIGONOMETRY

Unlike trigonometric identities that are true for all values of the defined variable, trigonometric equations are true for some, but not all, of the values of the variable. Most often trigonometric equations are solved for values between 0 and 360 degrees or 0 and 2π radians.

Some algebraic operation, such as squaring both sides of an equation, will give you extraneous answers. You must remember to check all solutions to be sure that they work.

Sample problems:

1. Solve: $\cos x = 1 - \sin x$ if $0 \leq x < 360$ degrees.

$\cos^2 x = (1 - \sin x)^2$ 1. square both sides

$1 - \sin^2 x = 1 - 2\sin x + \sin^2 x$ 2. substitute

$0 = {}^-2\sin x + 2\sin^2 x$ 3. set = to 0

$0 = 2\sin x({}^-1 + \sin x)$ 4. factor

$2\sin x = 0 \quad\quad {}^-1 + \sin x = 0$ 5. set each factor = 0

$\sin x = 0 \quad\quad \sin x = 1$ 6. solve for $\sin x$

$x = 0$ or $180 \quad\quad x = 90$ 7. find value of at x

The solutions appear to be 0, 90 and 180. Remember to check each solution and you will find that 180 does not give you a true equation. Therefore, the only solutions are 0 and 90 degrees.

2. Solve: $\cos^2 x = \sin^2 x$ if $0 \leq x < 2\pi$

$\cos^2 x = 1 - \cos^2 x$ 1. substitute

$2\cos^2 x = 1$ 2. simplify

$\cos^2 x = \dfrac{1}{2}$ 3. divide by 2

$\sqrt{\cos^2 x} = \pm\sqrt{\dfrac{1}{2}}$ 4. take square root

$\cos x = \dfrac{\pm\sqrt{2}}{2}$ 5. rationalize denominator

$x = \dfrac{\pi}{4}, \dfrac{3\pi}{4}, \dfrac{5\pi}{4}, \dfrac{7\pi}{4}$

Given 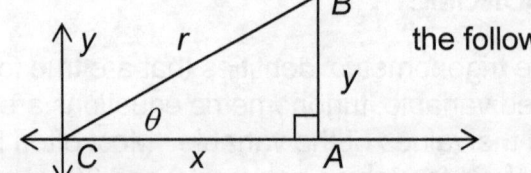 the following can be found.

Trigonometric Functions:

$\sin\theta = \dfrac{y}{r}$ $\qquad$ $\csc\theta = \dfrac{r}{y}$

$\cos\theta = \dfrac{x}{r}$ $\qquad$ $\sec\theta = \dfrac{r}{x}$

$\tan\theta = \dfrac{y}{x}$ $\qquad$ $\cot\theta = \dfrac{x}{y}$

Sample problem:

1. Prove that $\sec\theta = \dfrac{1}{\cos\theta}$.

$\sec\theta = \dfrac{1}{\frac{x}{r}}$ $\qquad$ Substitution definition of cosine.

$\sec\theta = \dfrac{1 \times r}{\frac{x}{r} \times r}$ $\qquad$ Multiply by $\dfrac{r}{r}$.

$\sec\theta = \dfrac{r}{x}$ $\qquad$ Substitution.

$\sec\theta = \sec\theta$ $\qquad$ Substitute definition of $\dfrac{r}{x}$.

$\sec\theta = \dfrac{1}{\cos\theta}$ $\qquad$ Substitute.

2. Prove that $\sin^2 + \cos^2 = 1$.

$\left(\dfrac{y}{r}\right)^2 + \left(\dfrac{x}{r}\right)^2 = 1$ $\qquad$ Substitute definitions of sin and cos.

$\dfrac{y^2 + x^2}{r^2} = 1$ $\qquad$ $x^2 + y^2 = r^2$ Pythagorean formula.

$\dfrac{r^2}{r^2} = 1$ $\qquad$ Simplify.

$1 = 1$ $\qquad$ Substitute.

$\sin^2\theta + \cos^2\theta = 1$

Practice problems: Prove each identity.

1. $\cot\theta = \dfrac{\cos\theta}{\sin\theta}$
2. $1+\cot^2\theta = \csc^2\theta$

There are two methods that may be used to prove trigonometric identities. One method is to choose one side of the equation and manipulate it until it equals the other side. The other method is to replace expressions on both sides of the equation with equivalent expressions until both sides are equal.

The Reciprocal Identities

$\sin x = \dfrac{1}{\csc x}$ $\qquad$ $\sin x \csc x = 1$ $\qquad$ $\csc x = \dfrac{1}{\sin x}$

$\cos x = \dfrac{1}{\sec x}$ $\qquad$ $\cos x \sec x = 1$ $\qquad$ $\sec x = \dfrac{1}{\cos x}$

$\tan x = \dfrac{1}{\cot x}$ $\qquad$ $\tan x \cot x = 1$ $\qquad$ $\cot x = \dfrac{1}{\tan x}$

$\tan x = \dfrac{\sin x}{\cos x}$ $\qquad\qquad\qquad\qquad\qquad$ $\cot x = \dfrac{\cos x}{\sin x}$

The Pythagorean Identities

$\sin^2 x + \cos^2 x = 1$ $\qquad$ $1+\tan^2 x = \sec^2 x$ $\qquad$ $1+\cot^2 x = \csc^2 x$

MATHEMATICS

Sample problems:

1. Prove that $\dfrac{\cos^2\theta}{1+2\sin\theta+\sin^2\theta} = \dfrac{\sec\theta - \tan\theta}{\sec\theta + \tan\theta}$.

$\dfrac{1-\sin^2\theta}{(1+\sin\theta)(1+\sin\theta)} = \dfrac{\sec\theta - \tan\theta}{\sec\theta + \tan\theta}$ Pythagorean identity

factor denominator.

$\dfrac{1-\sin^2\theta}{(1+\sin\theta)(1+\sin\theta)} = \dfrac{\dfrac{1}{\cos\theta} - \dfrac{\sin\theta}{\cos\theta}}{\dfrac{1}{\cos\theta} + \dfrac{\sin\theta}{\cos\theta}}$ Reciprocal

identities.

$\dfrac{(1-\sin\theta)(1+\sin\theta)}{(1+\sin\theta)(1+\sin\theta)} = \dfrac{\dfrac{1-\sin\theta}{\cos\theta}(\cos\theta)}{\dfrac{1+\sin\theta}{\cos\theta}(\cos\theta)}$ Factor $1-\sin^2\theta$.

Multiply by $\dfrac{\cos\theta}{\cos\theta}$.

$\dfrac{1-\sin\theta}{1+\sin\theta} = \dfrac{1-\sin\theta}{1+\sin\theta}$ Simplify.

$\dfrac{\cos^2\theta}{1+2\sin\theta+\sin^2\theta} = \dfrac{\sec\theta - \tan\theta}{\sec\theta + \tan\theta}$

Graph trigonometric functions.

It is easiest to graph trigonometric functions when using a calculator by making a table of values.

DEGREES

	0	30	45	60	90	120	135	150	180	210	225	240	270	300	315	330	360
sin	0	.5	.71	.87	1	.87	.71	.5	0	-.5	-.71	-.87	-1	-.87	-.71	-.5	0
cos	1	.87	.71	.5	0	-.5	-.71	-.87	-1	-.87	-.71	-.5	0	.5	.71	.87	1
tan	0	.57	1	1.7	--	-1.7	-1	-.57	0	.57	1	1.7	--	-1.7	-1	-.57	0

$0 \quad \dfrac{\pi}{6} \quad \dfrac{\pi}{4} \quad \dfrac{\pi}{3} \quad \dfrac{\pi}{2} \quad \dfrac{2\pi}{3} \quad \dfrac{3\pi}{4} \quad \dfrac{5\pi}{6} \quad \pi \quad \dfrac{7\pi}{6} \quad \dfrac{5\pi}{4} \quad \dfrac{4\pi}{3} \quad \dfrac{3\pi}{2} \quad \dfrac{5\pi}{3} \quad \dfrac{7\pi}{4} \quad \dfrac{11\pi}{6} \quad 2\pi$

RADIANS

Remember the graph always ranges from +1 to ⁻1 for sine and cosine functions unless noted as the coefficient of the function in the equation. For example, $y = 3\cos x$ has an amplitude of 3 units from the center line (0). Its maximum and minimum points would be at +3 and ⁻3.

Tangent is not defined at the values 90 and 270 degrees or $\dfrac{\pi}{2}$ and $\dfrac{3\pi}{2}$. Therefore, vertical asymptotes are drawn at those values.

The inverse functions can be graphed in the same manner using a calculator to create a table of values.

In order to solve **a right triangle** using trigonometric functions it is helpful to identify the given parts and label them. Usually more than one trigonometric function may be appropriately applied.

Some items to know about right triangles:

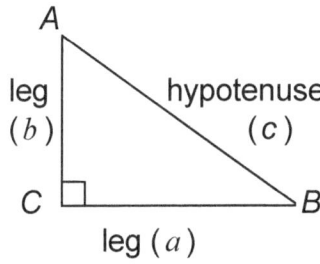

Given angle A, the side labeled leg (a) is adjacent angle A. And the side (b) is opposite to angle A

Sample problem:

1. Find the missing side.

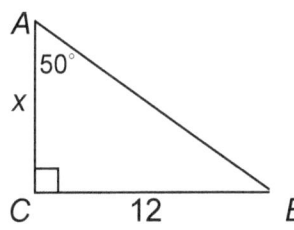

$\tan A = \dfrac{\text{opposite}}{\text{adjacent}}$

$\tan 50 = \dfrac{12}{x}$

$1.192 = \dfrac{12}{x}$

$x(1.192) = 12$

$x = 10.069$

1. Identify the known values. Angle $A = 50$ degrees. The side opposite the given angle is 12. The missing side is the adjacent leg.
2. The information suggests the use of the tangent function
3. Write the function.
4. Substitute.
5. Solve.

Remember that since angle A and angle B are complimentary, then angle $B = 90 - 50$ or 40 degrees.

Using this information we could have solved for the same side only this time it is the leg opposite from angle B.

$\tan B = \dfrac{\text{opposite}}{\text{adjacent}}$ 1. Write the formula.

$\tan 40 = \dfrac{x}{12}$ 2. Substitute.

$12(.839) = x$ 3. Solve.

$10.069 \approx x$

Now that the two sides of the triangle are known, the third side can be found using the Pythagorean Theorem.

Definition: For any triangle ABC, when given two sides and the included angle, the other side can be found using one of the formulas below:

$$a^2 = b^2 + c^2 - (2bc)\cos A$$
$$b^2 = a^2 + c^2 - (2ac)\cos B$$
$$c^2 = a^2 + b^2 - (2ab)\cos C$$

Similarly, when given three sides of a triangle, the included angles can be found using the derivation:

$$\cos A = \dfrac{b^2 + c^2 - a^2}{2bc}$$
$$\cos B = \dfrac{a^2 + c^2 - b^2}{2ac}$$
$$\cos C = \dfrac{a^2 + b^2 - c^2}{2ab}$$

Sample problem:

1. Solve triangle ABC, if angle $B = 87.5°$, $a = 12.3$, and $c = 23.2$. (Compute to the nearest tenth).

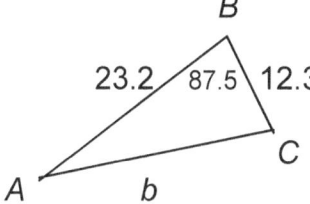

1. Draw and label a sketch.

Find side b.

$b^2 = a^2 + c^2 - (2ac)\cos B$

2. Write the formula.

$b^2 = (12.3)^2 + (23.2)^2 - 2(12.3)(23.2)(\cos 87.5)$

3. Substitute.

$b^2 = 664.636$

$b = 25.8$ (rounded)

4. Solve.

Use the law of sine to find angle A.

$\dfrac{\sin A}{a} = \dfrac{\sin B}{b}$

1. Write formula.

$\dfrac{\sin A}{12.3} = \dfrac{\sin 87.5}{25.8} = \dfrac{.999}{25.8}$

2. Substitute.

$\sin A = 0.47629$
Angle $A = 28.4$

3. Solve.

Therefore, angle $C = 180 - (87.5 + 28.4)$
$= 64.1$

2. Solve triangle *ABC* if $a = 15$, $b = 21$, and $c = 18$. (Round to the nearest tenth).

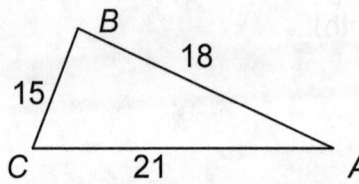

1. Draw and label a sketch.

Find angle *A*.

$\cos A = \dfrac{b^2 + c^2 - a^2}{2bc}$ 2. Write formula.

$\cos A = \dfrac{21^2 + 18^2 - 15^2}{2(21)(18)}$ 3. Substitute.

$\cos A = 0.714$ 4. Solve.
Angle $A = 44.4$

Find angle *B*.

$\cos B = \dfrac{a^2 + c^2 - b^2}{2ac}$ 5. Write formula.

$\cos B = \dfrac{15^2 + 18^2 - 21^2}{2(15)(18)}$ 6. Substitute.

$\cos B = 0.2$ 7. Solve.
Angle $B = 78.5$

Therefore, angle $C = 180 - (44.4 + 78.5)$
$ = 57.1$

Definition: For any triangle ABC, where a, b, and c are the lengths of the sides opposite angles A, B, and C respectively.

$$\frac{\sin A}{a} = \frac{\sin B}{b} = \frac{\sin C}{c}$$

Sample problem:

1. An inlet is 140 feet wide. The lines of sight from each bank to an approaching ship are 79 degrees and 58 degrees. What are the distances from each bank to the ship?

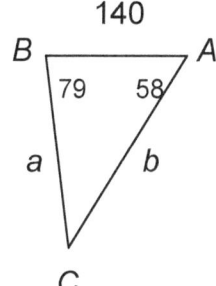

1. Draw and label a sketch
2. The missing angle is $180 - (79 + 58) = 43$.
   $\sphericalangle C = 43$ degrees

$$\frac{\sin A}{a} = \frac{\sin B}{b} = \frac{\sin C}{c}$$

3. Write formula.

Side opposite 79 degree angle:

$$\frac{\sin 79}{b} = \frac{\sin 43}{140}$$

$$b = \frac{140(.9816)}{.6820}$$

$b \approx 201.501$ feet

4. Substitute.

5. Solve.

Side opposite 58 degree angle:

$$\frac{\sin 58}{a} = \frac{\sin 43}{140}$$

$$a = \frac{140(.8480)}{.6820}$$

$a \approx 174.076$ feet

6. Substitute.

7. Solve.

When the measure of two sides and an angle not included between the sides are given, there are several possible solutions. There may be one, two or no triangles that may be formed from the given information.

The ambiguous case is described using two situations: either angle A is acute or it is obtuse.

Case 1: Angle A is acute.

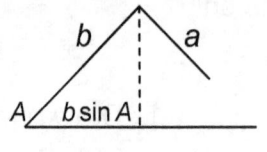
$a < b(\sin A)$
No triangle possible.

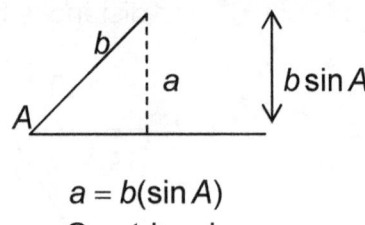
$a = b(\sin A)$
One triangle.

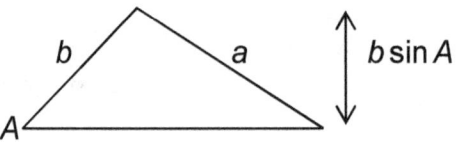
$a > b(\sin A)$ and $a \geq b$
One triangle.

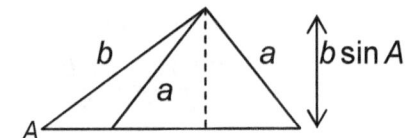
$a > b(\sin A)$ but $a < b$
Two triangles.

Case 2: Angle A is obtuse.

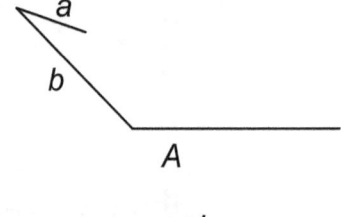
$a \leq b$
No triangle.

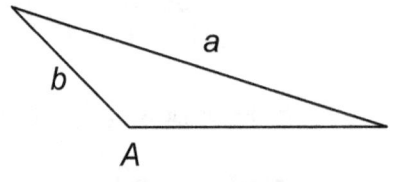
$a > b$
One triangle.

Sample problems:

Determine how many solutions exist.

1. Angle $A = 70$ degrees, $a = 16$, $b = 21$.
   Angle A is an acute angle and $a < b$ so we are looking at a case 1 triangle. In this case, $b(\sin A) = 19.734$. Since $a = 16$, this gives us a triangle possibility similar to case 1 triangle 1 where $a < b(\sin A)$ and there are no possible solutions.

2. Angle $C = 95.1$ degrees, $b = 16.8$, and $c = 10.9$.
   Angle C is an obtuse angle and $c \leq b$ (case 2, triangle 1). There are no possible solutions.

3. Angle $B = 45$ degrees, $b = 40$, and $c = 32$.
   Angle B is acute and $b \geq c$. Finding $c(\sin B)$ gives 22.627 and therefore, $b > c(\sin B)$. This indicates a case 1 triangle with one possible solution.

Find the number of possible solutions and then the missing sides, if possible (round all answers to the nearest whole number).

4. Angle $A = 37$ degrees, $a = 49$ and $b = 54$.
   Angle A is acute and $a < b$, find $b(\sin A)$. If $a > b(\sin A)$ there will be two triangles, and if $a < b(\sin A)$ there will be no triangles possible.

   $a = 49$ and $b(\sin A) = 32.498$, therefore, $a > b(\sin A)$ and there are two triangle solutions.

   Use law of sines to find angle B.

   $\dfrac{\sin A}{a} = \dfrac{\sin B}{b}$ 1. Write formula.

   $\dfrac{\sin 37}{49} = \dfrac{\sin B}{54}$ 2. Substitute.

   $\sin B = 0.663$ 3. Solve.

   Angle $B = 42$ degrees or angle $B = 180 - 42 = 138$ degrees. There are two possible solutions to be solved for.

Case 1 (angle $B = 42$ degrees)

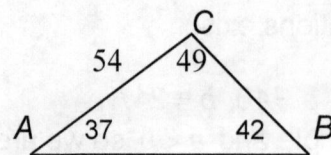

1. Draw sketch

Angle $C = 180 - (37 + 42)$
$ = 101$ degrees

2. Find angle C.

$$\frac{\sin A}{a} = \frac{\sin C}{c}$$

3. Find side c using law of sines

$$\frac{\sin 37}{49} = \frac{\sin 101}{c}$$
$c = 79.924 \approx 80$

4. Substitute.

Case 2 (angle $B = 138$)

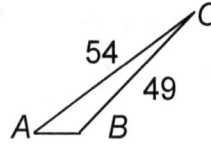

1. Draw a sketch. A=37 and B=138

Angle $C = 180 - (37 + 138)$
$ = 5$ degrees

2. Find angle C.

$$\frac{\sin A}{a} = \frac{\sin C}{c}$$

3. Find side c using law of sine.

$$\frac{\sin 37}{49} = \frac{\sin 5}{c}$$
$c = 7.097 \approx 7$

4. Substitute.

When finding the missing sides and angles of triangles that are either acute or obtuse using the law of sines and the law of cosines is imperative. Below is a chart to assist in determining the correct usage.

Given	Suggested Solution Method
(SAS) Two sides and the included angle	Law of Cosines will give you the third side. Then use the Law of Sines for the angles.
(SSS) Three sides	Law of Cosines will give you an angle. Then the Law of Sines can be used for the other angles.
(SAA or ASA) One side, two angles	Find the remaining angle and then use the Law of Sines.
(SSA) Two sides, angle not included	Find the number of possible solutions. Use the Law of Sines.

Definition: The area of any triangle *ABC* can be found using one of these formulas when given two legs and the included angle:

$$\text{Area} = \frac{1}{2} bc \sin A$$

$$\text{Area} = \frac{1}{2} ac \sin B$$

$$\text{Area} = \frac{1}{2} ab \sin C$$

Sample problem:

Find the area of triangle ABC with
a = 4.2, b = 2.6 and angle C = 43 degrees.

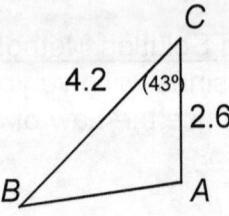

1. Draw and label the sketch.

$$\text{Area} = \frac{1}{2}ab\sin C$$

2. Write the formula.

$$= \frac{1}{2}(4.2)(2.6)\sin 43$$

3. Substitute.

$$= 3.724 \text{ square units}$$

4. Solve.

When only the lengths of the sides are known, it is possible to find the area of the triangle ABC using Heron's Formula:

$$\text{Area} = \sqrt{s(s-a)(s-b)(s-c)} \qquad \text{where } s = \frac{1}{2}(a+b+c)$$

Sample problem:

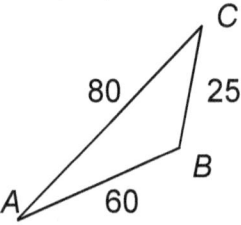

1. Draw and label the sketch.

First, find s.

$$s = \frac{1}{2}(a+b+c)$$

2. Write the formula.

$$s = \frac{1}{2}(25 + 80 + 60)$$

3. Substitute.

$$s = 82.5$$

4. Solve.

Now find the area.

$$A = \sqrt{s(s-a)(s-b)(s-c)}$$

5. Write the formula.

$$A = \sqrt{82.5(82.5-25)(82.5-80)(82.5-60)}$$

6. Substitute.

$$A = 516.5616 \text{ square units}$$

7. Solve.

One way to graph points is in the rectangular coordinate system. In this system, the point (a,b) describes the point whose distance along the x-axis is "a" and whose distance along the y-axis is "b." The other method used to locate points is the **polar plane coordinate system**. This system consists of a fixed point called the pole or origin (labeled O) and a ray with O as the initial point called the polar axis. The ordered pair of a point P in the polar coordinate system is (r,θ), where $|r|$ is the distance from the pole and θ is the angle measure from the polar axis to the ray formed by the pole and point P. The coordinates of the pole are $(0,\theta)$, where θ is arbitrary. Angle θ can be measured in either degrees or in radians.

Sample problem:

1. Graph the point P with polar coordinates $(^-2, ^-45 \text{ degrees})$.

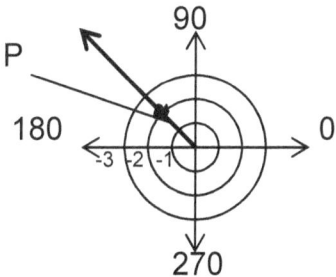

Draw $\theta = ^-45$ degrees in standard position. Since r is negative, locate the point $|^-2|$ units from the pole on the ray opposite the terminal side of the angle. Note that P can be represented by $(^-2, ^-45 \text{ degrees} + 180 \text{ degrees}) = (2, 135 \text{ degrees})$ or by $(^-2, ^-45 \text{ degrees} - 180 \text{ degrees}) = (2, ^-225 \text{ degrees})$.

2. Graph the point $P = \left(3, \dfrac{\pi}{4}\right)$ and show another graph that also represents the same point P.

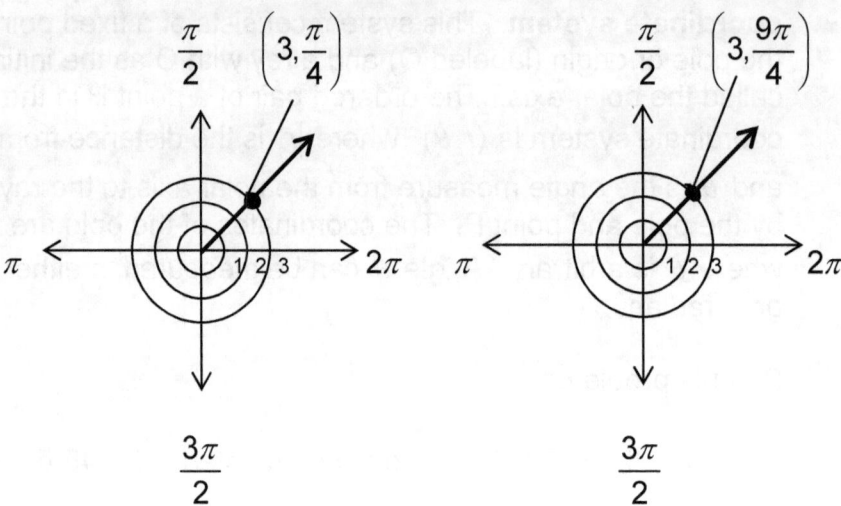

In the second graph, the angle 2π is added to $\dfrac{\pi}{4}$ to give the point $\left(3, \dfrac{9\pi}{4}\right)$.

It is possible that r is negative. Now instead of measuring $|r|$ units along the terminal side of the angle, we would locate the point $|^-3|$ units from the pole on the ray opposite the terminal side. This would give the points $\left(^-3, \dfrac{5\pi}{4}\right)$ and $\left(^-3, \dfrac{^-3\pi}{4}\right)$.

0005 FUNCTIONS

A function can be defined as a set of ordered pairs in which each element of the domain is paired with one and only one element of the range. The symbol $f(x)$ is read "f of x." A letter other than "f" can be used to represent a function. The letter "g" is commonly used as in $g(x)$.

Sample problems:

1. Given $f(x) = 4x^2 - 2x + 3$, find $f(^-3)$.

(This question is asking for the range value that corresponds to the domain value of $^-3$).

$f(x) = 4x^2 - 2x + 3$ 1. Replace x with -3

$f(^-3) = 4(^-3)^2 - 2(^-3) + 3$

$f(^-3) = 45$ 2. Solve.

2. Find f(3) and f(10), given $f(x) = 7$.

$f(x) = 7$ 1. There are no *x*
$(3) = 7$ values to substitute for.

$f(x) = 7$ This is your answer

$f10) = 7$ 2. Same as above.

Notice that both answers are equal to the constant given.

If $f(x)$ is a function and the value of 3 is in the domain, the corresponding element in the range would be f(3). It is found by evaluating the function for $x = 3$. The same holds true for adding, subtracting, and multiplying in function form.

The symbol f^{-1} is read "the inverse of f". The $^{-1}$ is not an exponent. The inverse of a function can be found by reversing the order of coordinates in each ordered pair that satisfies the function. Finding the inverse functions means switching the place of x and y and then solving for y.

Sample problem:

1. Find $p(a+1) + 3\{p(4a)\}$ if $p(x) = 2x^2 + x + 1$.

Find $p(a+1)$.

$$p(a+1) = 2(a+1)^2 + (a+1) + 1 \quad \text{Substitute } (a+1) \text{ for } x.$$
$$p(a+1) = 2a^2 + 5a + 4 \quad \text{Solve.}$$

Find $3\{p(4a)\}$.

$$3\{p(4a)\} = 3[2(4a)^2 + (4a) + 1] \quad \text{Substitute } (4a) \text{ for } x,$$
$$\text{multiply by 3.}$$
$$3\{p(4a)\} = 96a^2 + 12a + 3 \quad \text{Solve.}$$

$$p(a+1) + 3\{p(4a)\} = 2a^2 + 5a + 4 + 96a^2 + 12a + 3$$

Combine like terms.
$$p(a+1) + 3\{p(4a)\} = 98a^2 + 17a + 7$$

Definition: A function f is even if $f(^-x) = f(x)$ and odd if $f(^-x) = ^-f(x)$ for all x in the domain of f.

Sample problems:

Determine if the given function is even, odd, or neither even nor odd.

1. $f(x) = x^4 - 2x^2 + 7$
   $f(^-x) = (^-x)^4 - 2(^-x)^2 + 7$
   $f(^-x) = x^4 - 2x^2 + 7$

   1. Find $f(^-x)$.
   2. Replace x with ^-x.
   3. Since $f(^-x) = f(x)$, $f(x)$ is an even function.

   $f(x)$ is an even function.

2. $f(x) = 3x^3 + 2x$
   $f(^-x) = 3(^-x)^3 + 2(^-x)$
   $f(^-x) = ^-3x^3 - 2x$

   1. Find $f(^-x)$.
   2. Replace x with ^-x.
   3. Since $f(x)$ is not equal to $f(^-x)$, $f(x)$ is not an even function.

   $^-f(x) = ^-(3x^3 + 2x)$
   $^-f(x) = ^-3x^3 - 2x$

   4. Try $^-f(x)$.
   5. Since $f(^-x) = ^-f(x)$,

   $f(x)$ is an odd function.

3. $g(x) = 2x^2 - x + 4$
   $g(^-x) = 2(^-x)^2 - (^-x) + 4$
   $g(^-x) = 2x^2 + x + 4$

   1. First find $g(^-x)$.
   2. Replace x with ^-x.
   3. Since $g(x)$ does not equal $g(^-x)$, $g(x)$ is not an even function.

   $^-g(x) = ^-(2x^2 - x + 4)$
   $^-g(x) = ^-2x^2 + x - 4$

   4. Try $^-g(x)$.
   5. Since $^-g(x)$ does not equal $g(^-x)$, $g(x)$ is not an odd function.

   $g(x)$ is neither even nor odd.

A **family of functions** is a group of functions that share a certain attribute. For example, all functions with equations of the form $y = 3x + b$ belong to the $m = 3$ family and all functions with equations of the form $y = mx + 8$ belong to the $b = 8$ family.

Example:

An example of phenomena mentioned here could be population growth. When a population has a constant growth rate, we can calculate its size using a natural exponential function:

$$P(t) = P(0)e^{kt}$$

Where k is the constant relative growth rate, and $P(0)$ is the initial population at time zero.

Populations can also decrease. To represent this, we make the value of k negative and keep everything else the same:

$$P(t) = P(0)e^{-kt}$$

Sample problems:

Find the solution to the system of equations.

1. $y^2 - x^2 = {}^-9$
   $2y = x - 3$

   1. Use substitution method solving the second equation for x.

$2y = x - 3$
$x = 2y + 3$

2. Substitute this into the first equation in place of (x).

$y^2 - (2y+3)^2 = {}^-9$
$y^2 - (4y^2 + 12y + 9) = {}^-9$
$y^2 - 4y^2 - 12y - 9 = {}^-9$
${}^-3y^2 - 12y - 9 = {}^-9$
${}^-3y^2 - 12y = 0$
${}^-3y(y+4) = 0$

3. Solve.

4. Factor.

${}^-3y = 0 \quad y + 4 = 0$
$y = 0 \quad\quad y = {}^-4$

5. Set each factor equal to zero.
6. Use these values for y to solve for x.

$2y = x - 3 \quad\quad 2y = x - 3$
$2(0) = x - 3 \quad 2({}^-4) = x - 3$
$0 = x - 3 \quad\quad {}^-8 = x - 3$
$x = 3 \quad\quad\quad x = {}^-5$

7. Choose an equation.
8. Substitute.
9. Write ordered pairs.

$(3,0)$ and $({}^-5, {}^-4)$ satisfy the system of equations given.

MATHEMATICS

2. $-9x^2 + y^2 = 16$ Use elimination to solve.
   $5x^2 + y^2 = 30$

   $-9x^2 + y^2 = 16$
   $-5x^2 - y^2 = -30$

   1. Multiply second row by -1.
   $-14x^2 = -14$
   2. Add.
   $x^2 = 1$
   $x = \pm 1$
   3. Divide by -14. 4. Take the square root of both sides

   $-9(1)^2 + y^2 = 16$ $-9(-1)^2 + y^2 = 16$

   5. Substitute both values of x into the equation.
   $-9 + y^2 = 16$ $-9 + y^2 = 16$
   $y^2 = 25$ $y^2 = 25$
   6. Take the square root of both sides.
   $y = \pm 5$ $y = \pm 5$
   $(1, \pm 5)$ $(-1, \pm 5)$
   7. Write the ordered pairs.

   $(1, \pm 5)$ and $(-1, \pm 5)$ Satisfy the system of equations given.

A rational function is given in the form $f(x) = p(x)/q(x)$. In the equation, $p(x)$ and $q(x)$ both represent polynomial functions where $q(x)$ does not equal zero. The branches of rational functions approach asymptotes. Setting the denominator equal to zero and solving will give the value(s) of the vertical asymptotes(s) since the function will be undefined at this point. If the value of $f(x)$ approaches b as the $|x|$ increases, the equation $y = b$ is a horizontal asymptote. To find the horizontal asymptote it is necessary to make a table of values for x that are to the right and left of the vertical asymptotes. The pattern for the horizontal asymptotes will become apparent as the $|x|$ increases.

If there are more than one vertical asymptotes, remember to choose numbers to the right and left of each one in order to find the horizontal asymptotes and have sufficient points to graph the function.

Sample problem:

1. Graph $f(x) = \dfrac{3x+1}{x-2}$.

$x - 2 = 0$
$x = 2$

1. Set denominator $= 0$ to find the vertical asymptote.

x	f(x)
3	10
10	3.875
100	3.07
1000	3.007
1	⁻4
⁻10	2.417
⁻100	2.93
⁻1000	2.99

2. Make table choosing numbers to the right and left of the vertical asymptote.

3. The pattern shows that as the $|x|$ increases f(x) approaches the value 3, therefore a horizontal asymptote exists at $y = 3$

Sketch the graph.

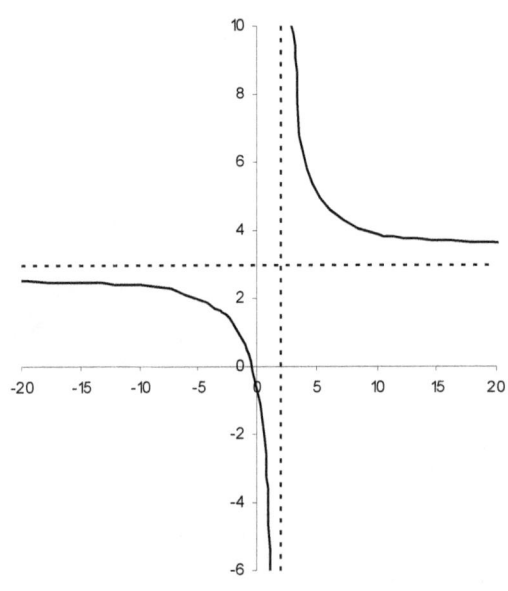

MATHEMATICS

When changing common logarithms to exponential form,

$$y = \log_b x \quad \text{if and only if} \quad x = b^y$$

Natural logarithms can be changed to exponential form by using,

$$\log_e x = \ln x \quad \text{or} \quad \ln x = y \quad \text{can be written as} \quad e^y = x$$

Practice Problems:

Express in exponential form.

1. $\log_3 81 = 4$
   $x = 81 \quad b = 3 \quad y = 4$ Identify values.
   $81 = 3^4$ Rewrite in exponential form.

Solve by writing in exponential form.

2. $\log_x 125 = 3$

   $x^3 = 125$ Write in exponential form.
   $x^3 = 5^3$ Write 125 in exponential form.
   $x = 5$ Bases must be equal if exponents are equal.

Use a scientific calculator to solve.

3. Find $\ln 72$.
   $\ln 72 = 4.2767$ Use the $\ln x$ key to find natural logs.

4. Find $\ln x = 4.2767$ Write in exponential form.
   $e^{4.2767} = x$ Use the key (or 2nd $\ln x$) to find
   $x = 72.002439$ x. The small difference is due to rounding.

To solve logarithms or exponential functions it is necessary to use several properties.

Multiplication Property $\quad\log_b mn = \log_b m + \log_b n$

Quotient Property $\quad\log_b \dfrac{m}{n} = \log_b m - \log_b n$

Powers Property $\quad\log_b n^r = r \log_b n$

Equality Property $\quad\log_b n = \log_b m$
if and only if $n = m$.

Change of Base Formula $\quad\log_b n = \dfrac{\log n}{\log b}$

$\log_b b^x = x$ and $b^{\log_b x} = x$

Sample problem.

Solve for x.

1. $\log_6(x-5) + \log_6 x = 2$

   $\log_6 x(x-5) = 2$ Use product property.

   $\log_6 x^2 - 5x = 2$ Distribute.

   $x^2 - 5x = 6^2$ Write in exponential form.

   $x^2 - 5x - 36 = 0$ Solve quadratic equation.

   $(x+4)(x-9) = 0$

   $x = {}^-4 \quad x = 9$

***Be sure to check results. Remember x must be greater than zero in $\log x = y$.

Check: $\log_6(x-5) + \log_6 x = 2$

$\log_6({}^-4 - 5) + \log_6({}^-4) = 2$ Substitute the first answer ${}^-4$.

$\log_6({}^-9) + \log_6({}^-4) = 2$ This is undefined, x is less than zero.

$\log_6(9-5) + \log_6 9 = 2$ Substitute the second answer 9.

$\log_6 4 + \log_6 9 = 2$

$\log_6(4)(9) = 2$ Multiplication property.

$\log_6 36 = 2$

$6^2 = 36$ Write in exponential form.

$36 = 36$

Practice problems:

1. $\log_4 x = 2\log_4 3$

2. $2\log_3 x = 2 + \log_3(x-2)$

3. Use change of base formula to find $(\log_3 4)(\log_4 3)$.

How to write the equation of the inverse of a function

1. To find the inverse of an equation using x and y, replace each letter with the other letter. Then solve the new equation for y, when possible. Given an equation like $y = 3x - 4$, replace each letter with the other:

   $x = 3y - 4$. Now solve this for y:
   $x + 4 = 3y$
   $1/3 x + 4/3 = y$ This is the inverse.

   Sometimes the function is named by a letter:

   $f(x) = 5x + 10$

   Temporarily replace f(x) with y.

   $y = 5x + 10$
   Now replace each letter with the other: $x = 5y + 10$
   Solve for the new y: $x - 10 = 5y$
   $1/5 x - 2 = y$

   The inverse of f(x) is denoted as $f^{-1}(x)$, so the answer is
   $f^{-1}(x) = 1/5 X - 2$.

- A first degree equation has an equation of the form $ax + by = c$. To graph this equation, find either one point and the slope of the line or find two points. To find a point and slope, solve the equation for y. This gets the equation in **slope intercept form**, $y = mx + b$. The point (0,b) is the y-intercept and m is the line's slope. To find any 2 points, substitute any 2 numbers for x and solve for y. To find the intercepts, substitute 0 for x and then 0 for y.
- Remember that graphs will go up as they go to the right when the slope is positive. Negative slopes make the lines go down as they go to the right.
- If the equation solves to *x* = **any number**, then the graph is a **vertical line**.
- If the equation solves to *y* = **any number**, then the graph is a **horizontal line**.
- When graphing a linear inequality, the line will be dotted if the inequality sign is < or >. If the inequality signs are either ≥ or ≤, the line on the graph will be a solid line. Shade above the line when the inequality sign is ≥ or >. Shade below the line when the inequality sign is < or ≤. Inequalities of the form $x >, x ≤, x <,$ or $x ≥$ number, draw a vertical line (solid or dotted). Shade to the right for > or ≥. Shade to the left for < or ≤. Remember:
Dividing or multiplying by a negative number will reverse the direction of the inequality sign.

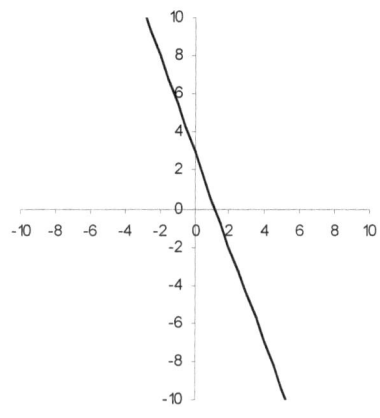
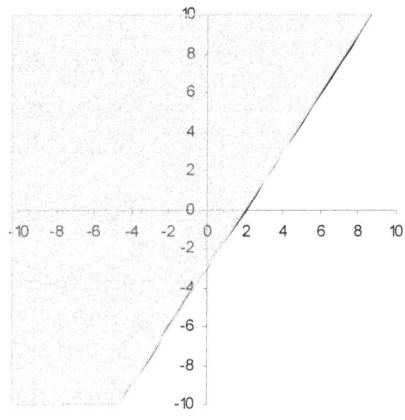

$5x + 2y = 6$

$y = {}^-5/2\, x + 3$

$3x - 2y \geq 6$

$y \leq 3/2\, x - 3$

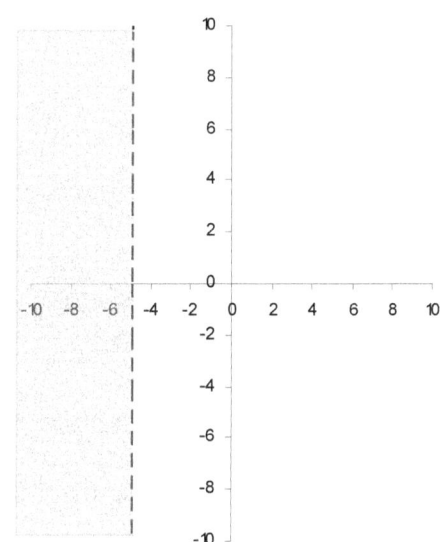

$3x + 12 < -3$

$x < {}^-5$

Graph the following:

1. $2x - y = {}^-4$
2. $x + 3y > 6$
3. $3x + 2y \leq 2y - 6$

0006   CALCULUS

The limit of a function is the *y* value that the graph approaches as the *x* values approach a certain number. To find a limit there are two points to remember.

1. Factor the expression completely and cancel all common factors in fractions.
2. Substitute the number to which the variable is approaching. In most cases this produces the value of the limit.

If the variable in the limit is approaching ∞, factor and simplify first; then examine the result. If the result does not involve a fraction with the variable in the denominator, the limit is usually also equal to ∞. If the variable is in the denominator of the fraction, the denominator is getting larger which makes the entire fraction smaller. In other words the limit is zero.

Examples:

1. $\lim\limits_{x \to {}^-3} \dfrac{x^2 + 5x + 6}{x + 3} + 4x$ Factor the numerator.

   $\lim\limits_{x \to {}^-3} \dfrac{(x + 3)(x + 2)}{(x + 3)} + 4x$ Cancel the common factors.

   $\lim\limits_{x \to {}^-3} (x + 2) + 4x$ Substitute $^-3$ for x.

   $({}^-3 + 2) + 4({}^-3)$ Simplify.

   $^-1 + {}^-12$

   $^-13$

2. $\lim\limits_{x \to \infty} \dfrac{2x^2}{x^5}$ Cancel the common factors.

   $\lim\limits_{x \to \infty} \dfrac{2}{x^3}$ Since the denominator is getting larger, the entire fraction is getting smaller.

   $\dfrac{2}{\infty^3}$

   0 The fraction is getting close to zero.

Practice problems:

1. $\lim_{x \to \pi} 5x^2 + \sin x$

2. $\lim_{x \to ^-4} \dfrac{x^2 + 9x + 20}{x + 4}$

After simplifying an expression to evaluate a limit, substitute the value that the variable approaches. If the substitution results in either $0/0$ or ∞/∞, use **L'Hopital's rule** to find the limit.

L'Hopital's rule states that you can find such limits by taking the derivative of the numerator and the derivative of the denominator, and then finding the limit of the resulting quotient.

Examples:

1. $\lim_{x \to \infty} \dfrac{3x - 1}{x^2 + 2x + 3}$ No factoring is possible.

$\dfrac{3\infty - 1}{\infty^2 + 2\infty + 3}$ Substitute ∞ for x.

$\dfrac{\infty}{\infty}$ Since a constant times infinity is still a large number, $3(\infty) = \infty$.

$\lim_{x \to \infty} \dfrac{3}{2x + 2}$ To find the limit, take the derivative of the numerator and denominator.

$\dfrac{3}{2(\infty) + 2}$ Substitute ∞ for x again.

$\dfrac{3}{\infty}$ Since the denominator is a very large number, the fraction is getting smaller.

$\dfrac{0}{1}$

Thus the limit is zero.

2. $\lim_{x \to 1} \dfrac{\ln x}{x-1}$ Substitute 1 for x.

$\dfrac{\ln 1}{1-1}$ The $\ln 1 = 0$

$\dfrac{0}{0}$ To find the limit, take the derivative of the numerator and denominator.

$\lim_{x \to 1} \dfrac{\frac{1}{x}}{1}$ Substitute 1 for x again.

$\dfrac{\frac{1}{1}}{1}$ Simplify. The limit is one.

1

Practice problems:

1. $\lim\limits_{x \to \infty} \dfrac{x^2 - 3}{x}$

2. $\lim\limits_{x \to \frac{\pi}{2}} \dfrac{\cos x}{x - \frac{\pi}{2}}$

To find the slope of a curve at a point, there are two steps to follow.

1. Take the derivative of the function.
2. Plug in the value to find the slope.

If plugging into the derivative yields a value of zero, the tangent line is horizontal at that point.

If plugging into the derivative produces a fraction with zero in the denominator, the tangent line at this point has an undefined slope and is thus a vertical line.

Examples:

1. Find the slope of the tangent line for the given function at the given point.

$$y = \frac{1}{x-2} \text{ at } (3,1)$$

$y = (x-2)^{-1}$ Rewrite using negative exponents.

$\frac{dy}{dx} = {}^-1(x-2)^{-1-1}(1)$ Use the Chain rule. The derivative of $(x-2)$ is 1.

$\frac{dy}{dx} = {}^-1(x-2)^{-2}$

$\frac{dy}{dx}\bigg|_{x=3} = {}^-1(3-2)^{-2}$ Evaluate at $x = 3$.

$\frac{dy}{dx}\bigg|_{x=3} = {}^-1$ The slope of the tangent line is $^-1$ at $x = 3$.

2. Find the points where the tangent to the curve $f(x) = 2x^2 + 3x$ is parallel to the line $y = 11x - 5$.

$f'(x) = 2 \bullet 2x^{2-1} + 3$ Take the derivative of $f(x)$ to get the slope of a tangent line.

$f'(x) = 4x + 3$

$4x + 3 = 11$ Set the slope expression $(4x + 3)$ equal to the slope of $y = 11x - 5$.

$x = 2$ Solve for the x value of the point.

$f(2) = 2(2)^2 + 3(2)$ The y value is 14.

$f(2) = 14$ So $(2,14)$ is the point on $f(x)$ where the tangent line is parallel to $y = 11x - 5$.

To write an equation of a tangent line at a point, two things are needed.

A point--the problem will usually provide a point, (x,y). If the problem only gives an x value, plug the value into the original function to get the y coordinate.

The slope--to find the slope, take the derivative of the original function. Then plug in the x value of the point to get the slope.

After obtaining a point and a slope, use the Point-Slope form for the equation of a line:

$$(y - y_1) = m(x - x_1)$$

where m is the slope and (x_1, y_1) is the point.

Example:

Find the equation of the tangent line to $f(x) = 2e^{x^2}$ at $x = {}^-1$.

$f({}^-1) = 2e^{({}^-1)^2}$	Plug in the x coordinate to obtain the y coordinate.
$= 2e^1$	The point is $({}^-1, 2e)$.
$f'(x) = 2e^{x^2} \bullet (2x)$	
$f'({}^-1) = 2e^{({}^-1)^2} \bullet (2 \bullet {}^-1)$	
$f'({}^-1) = 2e^1({}^-2)$	
$f'({}^-1) = {}^-4e$	The slope at $x = {}^-1$ is ${}^-4e$.
$(y - 2e) = {}^-4e(x - {}^-1)$	Plug in the point $({}^-1, 2e)$ and the slope $m = {}^-4e$.
$y = {}^-4ex - 4e + 2e$	Use the point slope form of a line.
$y = {}^-4ex - 2e$	Simplify to obtain the equation for the tangent line.

A normal line is a line which is perpendicular to a tangent line at a given point. Perpendicular lines have slopes which are negative reciprocals of each other. To find the equation of a normal line, first get the slope of the tangent line at the point. Find the negative reciprocal of this slope. Next, use the new slope and the point on the curve, both the x_1 and y_1 coordinates, and substitute into the Point-Slope form of the equation for a line:

$$(y - y_1) = slope \bullet (x - x_1)$$

Examples:

1. Find the equation of the normal line to the tangent to the curve $y = (x^2 - 1)(x - 3)$ at $x = {}^-2$.

$f(-2) = (({}^-2)^2 - 1)({}^-2 - 3)$ First find the y coordinate of the point on the curve. Here,

$$y = -\frac{1}{23}x - 14\frac{21}{23}$$

$y = {}^-15$ when $x = {}^-2$.

$y = x^3 - 3x^2 - x + 3$ Before taking the derivative, multiply the expression first. The derivative of a sum is easier to find than the derivative of a product.

$y' = 3x^2 - 6x - 1$ Take the derivative to find the slope of the tangent line.

$y'_{x={}^-2} = 3({}^-2)^2 - 6({}^-2) - 1$

$y'_{x={}^-2} = 23$

slope of normal $= \dfrac{{}^-1}{23}$

    For the slope of the normal line, take the negative reciprocal of the tangent line's slope.

$(y - {}^-15) = \dfrac{{}^-1}{23}(x - {}^-2)$

    Plug (x_1, y_1) into the point-slope equation.

$(y + 15) = \dfrac{{}^-1}{23}(x + 2)$

$y = -\dfrac{1}{23}x - 14\dfrac{21}{23}$

$y = -\dfrac{1}{23}x + \dfrac{2}{23} - 15 = \dfrac{1}{23}x - 14\dfrac{21}{23}$

MATHEMATICS

2. Find the equation of the normal line to the tangent to the curve $y = \ln(\sin x)$ at $x = \pi$.

$f(\pi) = \ln(\sin \pi)$

$f(\pi) = \ln(1) = 0$

$y' = \dfrac{1}{\sin x} \cdot \cos x$

$y'_{x=\pi} = \dfrac{\cos \pi}{\sin \pi} = \dfrac{0}{1}$

Slope of normal does not exist.

$\sin \pi = 1$ and $\ln(1) = 0$ (recall $e^0 = 1$).
So $x_1 = \pi$ and $y_1 = 0$.

Take the derivative to find the slope of the tangent line.

$\dfrac{-1}{0}$ does not exist. So the normal line is vertical at $x = \pi$.

A function is said to be increasing if it is rising from left to right and decreasing if it is falling from left to right. Lines with positive slopes are increasing, and lines with negative slopes are decreasing. If the function in question is something other than a line, simply refer to the slopes of the tangent lines as the test for increasing or decreasing. Take the derivative of the function and plug in an *x* value to get the slope of the tangent line; a positive slope means the function is increasing and a negative slope means it is decreasing. If an interval for *x* values is given, just pick any point between the two values to substitute.

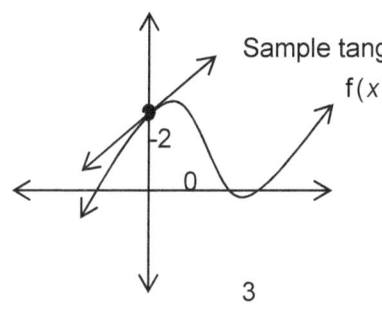

Sample tangent line on $(^-2,0)$

On the interval $(^-2,0)$, $f(x)$ is increasing. The tangent lines on this part of the graph have positive slopes.

Example:

The growth of a certain bacteria is given by $f(x) = x + \dfrac{1}{x}$. Determine if the rate of growth is increasing or decreasing on the time interval $(^-1,0)$.

$f'(x) = 1 + \dfrac{^-1}{x^2}$

To test for increasing or decreasing, find the slope of the tangent line by taking the derivative.

$f'\left(\dfrac{^-1}{2}\right) = 1 + \dfrac{^-1}{(^-1/2)^2}$

Pick any point on $(^-1,0)$ and substitute into the derivative.

$f'\left(\dfrac{^-1}{2}\right) = 1 + \dfrac{^-1}{1/4}$

$= 1 - 4$

$= ^-3$

The slope of the tangent line at $x = \dfrac{^-1}{2}$ is $^-3$.

The exact value of the slope is not important. The important fact is that the slope is negative.

Substituting an x value into a function produces a corresponding y value. The coordinates of the point (x,y), where y is the largest of all the y values, is said to be a maximum point. The coordinates of the point (x,y), where y is the smallest of all the y values, is said to be a minimum point. To find these points, only a few x values must be tested. First, find all of the x values that make the derivative either zero or undefined. Substitute these values into the original function to obtain the corresponding y values. Compare the y values. The largest y value is a maximum; the smallest y value is a minimum. If the question asks for the maxima or minima on an interval, be certain to also find the y values that correspond to the numbers at either end of the interval.

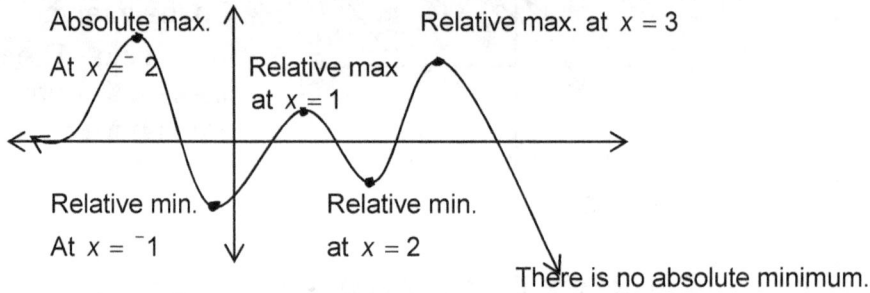

Example:
Find the maxima and minima of $f(x) = 2x^4 - 4x^2$ at the interval $(^-2, 1)$.

$f'(x) = 8x^3 - 8x$	Take the derivative first. Find all
$8x^3 - 8x = 0$	the x values (critical values) that
$8x(x^2 - 1) = 0$	make the derivative zero or
$8x(x - 1)(x + 1) = 0$	undefined. In this case, there are
$x = 0,\ x = 1,\ \text{or}\ x = {}^-1$	no x values that make the derivative undefined.
$f(0) = 2(0)^4 - 4(0)^2 = 0$	Substitute the critical values into the original function.
$f(1) = 2(1)^4 - 4(1)^2 = {}^-2$	Also, plug in the endpoint of
$f(^-1) = 2(^-1)^4 - 4(^-1)^2 = {}^-2$	the interval. Note that 1 is
$f(^-2) = 2(^-2)^4 - 4(^-2)^2 = 16$	a critical point and an endpoint.

The maximum is at (-2, 16) and there are minima at (1, -2) and (-1, -2). (0,0) is neither the maximum or minimum on (-2, 1) but it is still considered a relative extra point.

The first derivative reveals whether a curve is rising or falling (increasing or decreasing) from the left to the right. In much the same way, the second derivative relates whether the curve is concave up or concave down. Curves which are concave up are said to "collect water;" curves which are concave down are said to "dump water." To find the intervals where a curve is concave up or concave down, follow the following steps.

1. Take the second derivative (i.e. the derivative of the first derivative).
2. Find the critical x values.
   -Set the second derivative equal to zero and solve for critical x values.
   -Find the x values that make the second derivative undefined (i.e. make the denominator of the second derivative equal to zero). Such values may not always exist.
3. Pick sample values which are both less than and greater than each of the critical values.
4. Substitute each of these sample values into the second derivative and determine whether the result is positive or negative.

-If the sample value yields a positive number for the second derivative, the curve is concave up on the interval where the sample value originated.

-If the sample value yields a negative number for the second derivative, the curve is concave down on the interval where the sample value originated.

Example:

Find the intervals where the curve is concave up and concave down for $f(x) = x^4 - 4x^3 + 16x - 16$.

$f'(x) = 4x^3 - 12x^2 + 16$ — Take the second derivative.

$f''(x) = 12x^2 - 24x$ — Find the critical values by setting the second derivative equal to zero.

$12x^2 - 24x = 0$
$12x(x - 2) = 0$

There are no values that make the second derivative undefined.

$x = 0$ or $x = 2$

Set up a number line with the critical values.

Sample values: $^-1, 1, 3$ — Pick sample values in each of the 3 intervals.

$f''(^-1) = 12(^-1)^2 - 24(^-1) = 36$

$f''(1) = 12(1)^2 - 24(1) = {}^-12$

$f''(3) = 12(3)^2 - 24(3) = 36$

If the sample value produces a negative number, the function is concave down.

If the value produces a positive number, the curve is concave up. If the value produces a zero, the function is linear.

Therefore when $x < 0$ the function is concave up,
when $0 < x < 2$ the function is concave down,
when $x > 2$ the function is concave up.

A point of inflection is a point where a curve changes from being concave up to concave down or vice versa. To find these points, follow the steps for finding the intervals where a curve is concave up or concave down. A critical value is part of an inflection point if the curve is concave up on one side of the value and concave down on the other. The critical value is the x coordinate of the inflection point. To get the y coordinate, plug the critical value into the **original** function.

Example: Find the inflection points of $f(x) = 2x - \tan x$ where $\dfrac{-\pi}{2} < x < \dfrac{\pi}{2}$.

$(x) = 2x - \tan x \qquad \dfrac{-\pi}{2} < x < \dfrac{\pi}{2}$

Note the restriction on x.

$f'(x) = 2 - \sec^2 x$

Take the second derivative. Use the Power rule.

$f''(x) = 0 - 2 \bullet \sec x \bullet (\sec x \tan x)$

$= {}^-2 \bullet \dfrac{1}{\cos x} \bullet \dfrac{1}{\cos x} \bullet \dfrac{\sin x}{\cos x}$

The derivative of $\sec x$ is $(\sec x \tan x)$.

$f''(x) = \dfrac{{}^-2\sin x}{\cos^3 x}$

Find critical values by solving for the second derivative equal to zero.

$0 = \dfrac{{}^-2\sin x}{\cos^3 x}$

No x values on $\left(\dfrac{-\pi}{2}, \dfrac{\pi}{2}\right)$ make the denominator zero.

${}^-2\sin x = 0$
$\sin x = 0$
$x = 0$

Pick sample values on each side of the critical value $x = 0$.

Sample values: $x = \frac{-\pi}{4}$ and $x = \frac{\pi}{4}$

$$f''\left(\frac{-\pi}{4}\right) = \frac{-2\sin(-\pi/4)}{\cos^3(-\pi/4)} = \frac{-2(-\sqrt{2}/2)}{(\sqrt{2}/2)^3} = \frac{\sqrt{2}}{(\sqrt{8}/8)} = \frac{8\sqrt{2}}{\sqrt{8}} = \frac{8\sqrt{2}}{\sqrt{8}} \cdot \frac{\sqrt{8}}{\sqrt{8}}$$

$$= \frac{8\sqrt{16}}{8} = 4$$

$$f''\left(\frac{\pi}{4}\right) = \frac{-2\sin(\pi/4)}{\cos^3(\pi/4)} = \frac{-2(\sqrt{2}/2)}{(\sqrt{2}/2)^3} = \frac{-\sqrt{2}}{(\sqrt{8}/8)} = \frac{-8\sqrt{2}}{\sqrt{8}} = -4$$

The second derivative is positive on $(0, \infty)$ and negative on $(-\infty, 0)$. So the curve changes concavity at $x = 0$. se the original equation to find the y value that inflection occurs at.

$f(0) = 2(0) - \tan 0 = 0 - 0 = 0$ The inflection point is (0,0).

Extreme value problems are also known as max-min problems. Extreme value problems require using the first derivative to find values which either maximize or minimize some quantity such as area, profit, or volume. Follow these steps to solve an extreme value problem.

1. Write an equation for the quantity to be maximized or minimized.
2. Use the other information in the problem to write secondary equations.
3. Use the secondary equations for substitutions, and rewrite the original equation in terms of only one variable.
4. Find the derivative of the primary equation (step 1) and the critical values of this derivative.
5. Substitute these critical values into the primary equation.

The value which produces either the largest or smallest value is used to find the solution.

Example:

A manufacturer wishes to construct an open box from the piece of metal shown below by cutting squares from each corner and folding up the sides. The square piece of metal is 12 feet on a side. What are the dimensions of the squares to be cut out which will maximize the volume?

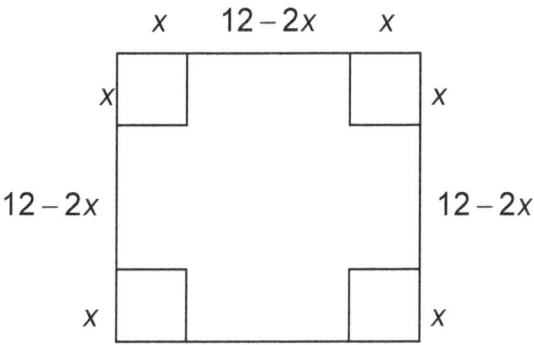

Volume = lwh Primary equation.
$l = 12 - 2x$
$w = 12 - 2x$ Secondary equations.
$h = x$
$V = (12 - 2x)(12 - 2x)(x)$ Make substitutions.
$V = (144x - 48x^2 + 4x^3)$ Take the derivative.
$\dfrac{dV}{dx} = 144 - 96x + 12x^2$

$0 = 12(x^2 - 8x + 12)$ Find critical values by setting the derivative equal to zero.
$0 = 12(x - 6)(x - 2)$
$x = 6$ and $x = 2$ Substitute critical values into volume equation.

$V = 144(6) - 48(6)^2 + 4(6)^3$ $V = 144(2) - 48(2)^2 + 4(2)^3$
$V = 0$ ft^3 when $x = 6$ $V = 128$ ft^3 when $x = 2$

Therefore, the manufacturer can maximize the volume if the squares to be cut out are 2 feet by 2 feet ($x = 2$).

If a particle (or a car, a bullet, etc.) is moving along a line, then the distance that the particle travels can be expressed by a function in terms of time.

1. The first derivative of the distance function will provide the velocity function for the particle. Substituting a value for time into this expression will provide the instantaneous velocity of the particle at the time. Velocity is the rate of change of the distance traveled by the particle. Taking the absolute value of the derivative provides the speed of the particle. A positive value for the velocity indicates that the particle is moving forward, and a negative value indicates the particle is moving backwards.

2. The second derivative of the distance function (which would also be the first derivative of the velocity function) provides the acceleration function. The acceleration of the particle is the rate of change of the velocity. If a value for time produces a positive acceleration, the particle is speeding up; if it produces a negative value, the particle is slowing down. If the acceleration is zero, the particle is moving at a constant speed.

To find the time when a particle stops, set the first derivative (i.e. the velocity function) equal to zero and solve for time. This time value is also the instant when the particle changes direction.

Example:

The motion of a particle moving along a line is according to the equation:

$s(t) = 20 + 3t - 5t^2$ where s is in meters and t is in seconds. Find the position, velocity, and acceleration of a particle at $t = 2$ seconds.

$s(2) = 20 + 3(2) - 5(2)^2$ $ = 6$ meters	Plug $t = 2$ into the original equation to find the position.
$s\,'(t) = v(t) = 3 - 10t$	The derivative of the first function gives the velocity.
$v(2) = 3 - 10(2) = {}^-17$ m/s	Plug $t = 2$ into the velocity function to find the velocity. ${}^-17$ m/s indicates the particle is moving backwards.
$s\,''(t) = a(t) = {}^-10$	The second derivation of position gives the acceleration.

$a(2) = {}^-10\, m/s^2$

Substitute $t=2$, yields an acceleration of $^-10\, m/s^2$, which indicates the particle is slowing down.

Finding the rate of change of one quantity (for example distance, volume, etc.) with respect to time it is often referred to as a rate of change problem. To find an instantaneous rate of change of a particular quantity, write a function in terms of time for that quantity; then take the derivative of the function. Substitute in the values at which the instantaneous rate of change is sought.

Functions which are in terms of more than one variable may be used to find related rates of change. These functions are often not written in terms of time. To find a related rate of change, follow these steps.

1. Write an equation which relates all the quantities referred to in the problem.
2. Take the derivative of both sides of the equation with respect to time. Follow the same steps as used in implicit differentiation. This means take the derivative of each part of the equation remembering to multiply each term by the derivative of the variable involved with respect to time. For example, if a term includes the variable v for volume, take the derivative of the term remembering to multiply by dv/dt for the derivative of volume with respect to time. dv/dt is the rate of change of the volume.
3. Substitute the known rates of change and quantities, and solve for the desired rate of change.

Example:

1. What is the instantaneous rate of change of the area of a circle where the radius is 3 cm?

$A(r) = \pi r^2$ Write an equation for area.
$A'(r) = 2\pi r$ Take the derivative to find the rate of change.
$A'(3) = 2\pi(3) = 6\pi$ Substitute in $r=3$ to arrive at the instantaneous rate of change.

Taking the antiderivative of a function is the opposite of taking the derivative of the function--much in the same way that squaring an expression is the opposite of taking the square root of the expression. For example, since the derivative of x^2 is $2x$ then the antiderivative of $2x$ is x^2. The key to being able to take antiderivatives is being as familiar as possible with the derivative rules.

To take the antiderivative of an algebraic function (the sum of products of coefficients and variables raised to powers other than negative one), take the antiderivative of each term in the function by following these steps.

1. Take the antiderivative of each term separately.
2. The coefficient of the variable should be equal to one plus the exponent.
3. If the coefficient is not one more than the exponent, put the correct coefficient on the variable and also multiply by the reciprocal of the number put in.
   Ex. For $4x^5$, the coefficient should be 6 not 4. So put in 6 and the reciprocal 1/6 to achieve $(4/6)6x^5$.
4. Finally take the antiderivative by replacing the coefficient and variable with just the variable raised to one plus the original exponent.
   Ex. For $(4/6)6x^5$, the antiderivative is $(4/6)x^6$.
   You have to add in constant c because there is no way to know if a constant was originally present since the derivative of a constant is zero.
5. Check your work by taking the first derivative of your answer. You should get the original algebraic function.

Examples: Take the antiderivative of each function.

1. $f(x) = 5x^4 + 2x$ The coefficient of each term is already 1 more than the exponent.
   $F(x) = x^5 + x^2 + c$ $F(x)$ is the antiderivative of $f(x)$
   $F'(x) = 5x^4 + 2x$ Check by taking the derivative of $F(x)$.

2. $f(x) = {}^-2x^{-3}$
   $F(x) = x^{-2} + c = \dfrac{1}{x^2} + c$ $F(x)$ is the antiderivative of $f(x)$.
   $F'(x) = {}^-2x^{-3}$ Check.

3. $f(x) = {}^-4x^2 + 2x^7$ Neither coefficient is correct.

$f(x) = {}^-4 \cdot \dfrac{1}{3} \cdot 3x^2 + 2 \cdot \dfrac{1}{8} \cdot 8x^7$

Put in the correct coefficient along with its reciprocal.

$F(x) = {}^-4 \cdot \dfrac{1}{3} x^3 + 2 \cdot \dfrac{1}{8} x^8 + c$

$F(x)$ is the antiderivative of $f(x)$.

$F(x) = \dfrac{{}^-4}{3} x^3 + \dfrac{1}{4} x^8 + c$

$F'(x) = {}^-4x^2 + 2x^7$ Check.

The rules for taking antiderivatives of trigonometric functions follow, but be aware that these can get very confusing to memorize because they are very similar to the derivative rules. Check the antiderivative you get by taking the derivative and comparing it to the original function.

1. $\sin x$ the antiderivative for $\sin x$ is ${}^-\cos x + c$.
2. $\cos x$ the antiderivative for $\cos x$ is $\sin x + c$.
3. $\tan x$ the antiderivative for $\tan x$ is $-\ln|\cos x| + c$.
4. $\sec^2 x$ the antiderivative for $\sec^2 x$ is $\tan x + c$.
5. $\sec x \tan x$ the antiderivative for $\sec x \tan x$ is $\sec x + c$.

If the trigonometric function has a coefficient, simply keep the coefficient and multiply the antiderivative by it.

Examples: Find the antiderivatives for the following functions.

1. $f(x) = 2 \sin x$ Carry the 2 throughout the problem.

$F(x) = 2({}^-\cos x) = {}^-2\cos x + c$ $F(x)$ is the antiderivative.

$F'(x) = {}^-2({}^-\sin x) = 2\sin x$ Check by taking the derivative of $F(x)$.

MATHEMATICS

2. $f(x) = \dfrac{\tan x}{5}$

$F(x) = \dfrac{-\ln|\cos x|}{5} + c$ $F(x)$ is the antiderivative of $f(x)$.

$F'(x) = \dfrac{1}{5}\left(-\dfrac{1}{|\cos x|}\right)(-\sin x) = \dfrac{1}{5}\tan x$ Check by taking the derivative of $F(x)$.

Practice problems: Find the antiderivative of each function.

1. $f(x) = {}^-20\cos x$ 2. $f(x) = \pi \sec x \tan x$

Use the following rules when finding the antiderivative of an exponential function.

1. e^x The antiderivative of e^x is the same $e^x + c$.
2. a^x The antiderivative of a^x, where a is any number, is $a^x / \ln a + c$.

Examples: Find the antiderivatives of the following functions:

1. $f(x) = 10e^x$

    $F(x) = 10e^x + c$ $F(x)$ is the antiderivative.

    $F'(x) = 10e^x$ Check by taking the derivative of $F(x)$.

2. $f(x) = \dfrac{2^x}{3}$

    $F(x) = \dfrac{1}{3} \cdot \dfrac{2^x}{\ln 2} + c$ $F(x)$ is the antiderivative.

    $F'(x) = \dfrac{1}{3\ln 2}\ln 2 (2^x)$ Check by taking the derivative of $F(x)$.

    $F'(x) = \dfrac{2^x}{3}$

The derivative of a distance function provides a velocity function, and the derivative of a velocity function provides an acceleration function. Therefore taking the antiderivative of an acceleration function yields the velocity function, and the antiderivative of the velocity function yields the distance function.

Example:

A particle moves along the x axis with acceleration $a(t) = 3t - 1$ cm/sec/sec. At time $t = 4$, the particle is moving to the left at 3 cm per second. Find the velocity of the particle at time $t = 2$ seconds.

$a(t) = 3t - 1$ Before taking the antiderivative, make sure the correct coefficients are present.

$a(t) = 3 \cdot \dfrac{1}{2} \cdot 2t - 1$

$v(t) = \dfrac{3}{2}t^2 - 1 \cdot t + c$ $v(t)$ is the antiderivative of $a(t)$.

$v(4) = \dfrac{3}{2}(4)^2 - 1(4) + c = {}^-3$

$v(4) = {}^-3$ Use the given information that to find c.

$24 - 4 + c = {}^- 3$

$20 + c = {}^-3$

$c = {}^-23$ The constant is ${}^-23$.

$v(t) = \dfrac{3}{2}t^2 - 1t + {}^-23$ Rewrite $v(t)$ using $c = {}^-23$.

$v(2) = \dfrac{3}{2}2^2 - 1(2) + {}^-23$ Solve $v(t)$ at $t=2$.

$v(2) = 6 + 2 + {}^-23 = {}^-15$ The velocity at $t = 2$ is -15 cm/sec.

Practice problem:

A particle moves along a line with acceleration $a(t) = 5t + 2$. The velocity after 2 seconds is -10m/sec.

1. Find the initial velocity.
2. Find the velocity at $t = 4$.

To find the distance function, take the antiderivative of the velocity function. And to find the velocity function, find the antiderivative of the acceleration function. Use the information in the problem to solve for the constants that result from taking the antiderivatives.

Example:

A particle moves along the x axis with acceleration $a(t) = 6t - 6$. The initial velocity is 0 m/sec and the initial position is 8 cm to the right of the origin. Find the velocity and position functions.

$v(0) = 0$
$s(0) = 8$
$a(t) = 6t - 6$ Interpret the given information.

Put in the coefficients needed to take the antiderivative.

$a(t) = 6 \cdot \dfrac{1}{2} \cdot 2t - 6$

$v(t) = \dfrac{6}{2}t^2 - 6t + c$ Take the antiderivative of $a(t)$ to get $v(t)$.

$vv(0) = 3(0)^2 - 6(0) = c = 0$ Use $v(0) = 0$ to solve for c.
$0 - 0 + c = 0$
$c = 0$ $c = 0$

$v(t) = 3t^2 - 6t + 0$ Rewrite $v(t)$ using $c = 0$.

$v(t) = 3t^2 - 6\dfrac{1}{2} \cdot 2t$

Put in the coefficients needed to take the antiderivative.

$s(t) = t^3 - \dfrac{6}{2}t^2 + c$ Take the antiderivative of $v(t)$ to get $s(t)$ → the distance function.

$s(0) = 0^3 - 3(0)^2 + c = 8$ Use $s(0) = 8$ to solve for c.
$c = 8$
$s(t) = t^3 - 3t^2 + 8$

An integral is almost the same thing as an antiderivative, the only difference is the notation.

$\int_{-2}^{1} 2x\,dx$ is the integral form of the antiderivative of $2x$. The numbers at the top and bottom of the integral sign (1 and $^-2$) are the numbers used to find the exact value of this integral. If these numbers are used the integral is said to be *definite* and does not have an unknown constant c in the answer.

The fundamental theorem of calculus states that an integral such as the one above is equal to the antiderivative of the function inside (here $2x$) evaluated from $x = {}^-2$ to $x = 1$. To do this, follow these steps.

1. Take the antiderivative of the function inside the integral.
2. Plug in the upper number (here $x = 1$) and plug in the lower number (here $x = {}^-2$), giving two expressions.
3. Subtract the second expression from the first to achieve the integral value.

Examples:

1. $\int_{-2}^{1} 2x\,dx = x^2 \big]_{-2}^{1}$ Take the antiderivative of.

   $\int_{-2}^{1} 2x\,dx = 1^2 - (^-2)^2$ Substitute in $x = 1$ and $x = {}^-2$ and subtract the results.

   $\int_{-2}^{1} 2x\,dx = 1 - 4 = {}^-3$ The integral has the value $^-3$.

2. $\int_{0}^{\pi/2} \cos x\,dx = \sin x \big]_{0}^{\pi/2}$ The antiderivative of $\cos x$ is $\sin x$.

   $\int_{0}^{\pi/2} \cos x\,dx = \sin\frac{\pi}{2} - \sin 0$ Substitute in $x = \frac{\pi}{2}$ and $x = 0$. Subtract the results.

   $\int_{0}^{\pi/2} \cos x\,dx = 1 - 0 = 1$ The integral has the value 1.

A list of integration formulas follows. In each case the letter u is used to represent either a single variable or an expression. Note that also in each case du is required. du is the derivative of whatever u stands for. If u is sin x then du is cos x, which is the derivative of sin x.

If the derivative of u is not entirely present, remember you can put in constants as long as you also insert the reciprocal of any such constants. n is a natural number.

$$\int u^n du = \frac{1}{n+1} u^{n+1} + c \text{ if } n \neq {}^-1$$

$$\int \frac{1}{u} du = \ln|u| + c$$

$$\int e^u du = e^u + c$$

$$\int a^u du = \frac{1}{\ln a} a^u + c$$

$$\int \sin u \, du = {}^- \cos u + c$$

$$\int \cos u \, du = \sin u + c$$

$$\int \sec^2 u \, du = \tan u + c$$

$$\int \csc^2 u \, du = {}^- \cot u + c$$

Example:

1. $\int \frac{6}{x} dx = 6 \int \frac{1}{x} dx$ You can pull any constants outside the integral.

   $\int \frac{6}{x} dx = 6 \ln|x| + c$

Sometimes an **integral** does not always look exactly like the forms above. But with a simple substitution (sometimes called a u substitution), the integral can be made to look like one of the general forms.

You might need to experiment with different u substitutions before you find the one that works. Follow these steps.

1. If the object of the integral is a sum or difference, first split the integral up.
2. For each integral, see if it fits one of the general forms from the previous page.
3. If the integral does not fit one of the forms, substitute the letter u in place of one of the expressions in the integral.

4. Off to the side, take the derivative of *u*, and see if that derivative exists inside the original integral. If it does, replace that derivative expression with *du*. If it does not, try another *u* substitution.
5. Now the integral should match one of the general forms, including both the *u* and the *du*.
6. Take the integral using the general forms, and substitute for the value of *u*.

Examples:

1. $\int (\sin x^2 \cdot 2x + \cos x^2 \cdot 2x) dx$ Split the integral up.

   $\int \sin(x^2) \cdot 2x\, dx + \int \cos(x^2) 2x\, dx$

   $u = x^2, \quad du = 2x\, dx$ If you let $u = x^2$, the derivative of *u*, *du*, is $2x\,dx$.

   $\int \sin u\, du + \int \cos u\, du$

   Make the *u* and *du* substitutions.

   $^-\cos u + \sin u + c$

   Use the formula for integrating $\sin u$ and $\cos u$.

   $^-\cos(x^2) + \sin(x^2) + c$ Substitute back in for *u*.

2. $\int e^{\sin x} \cos x\, dx$

   Try letting $u = \cos x$.

   $u = \cos x, \; du = {}^-\sin x\, dx$ The derivative of *u* would be $^-\sin x\, dx$, which is not present.

   $u = \sin x, \; du = \cos x\, dx$ Try another substitution: $u = \sin x$, $du = \cos x\, dx$. $du = \cos x\, dx$ is present.

   $\int e^u\, du$ e^u is one of the general forms.

   e^u The integral of e^u is e^u.

   $e^{\sin x} + C$ Substitute back in for *u*.

MATHEMATICS

TEACHER CERTIFICATION STUDY GUIDE

Integration by parts should only be used if none of the other integration methods works. Integration by parts requires two substitutions (both *u* and *dv*).

1. Let *dv* be the part of the integral that you think can be integrated by itself.
2. Let *u* be the part of the integral which is left after the *dv* substitution is made.
3. Integrate the *dv* expression to arrive at just simply *v*.
4. Differentiate the *u* expression to arrive at *du*. If *u* is just *x*, then *du* is dx.
5. Rewrite the integral using $\int u\,dv = uv - \int v\,du$.
6. All that is left is to integrate $\int v\,du$.
7. If you cannot integrate *v du*, try a different set of substitutions and start the process over.

Examples:

1. $\int xe^{3x}\,dx$ Make *dv* and *u* substitutions.

$dv = e^{3x}\,dx \quad u = x$ Integrate the *dv* term to arrive at *v*.

$v = \frac{1}{3}e^{3x} \quad du = dx$ Differentiate the *u* term to arrive at *du*.

$\int xe^{3x}\,dx = x\left(\frac{1}{3}e^{3x}\right) - \int \frac{1}{3}e^{3x}\,dx$ Rewrite the integral using the above formula.

$\int xe^{3x}\,dx = \frac{1}{3}xe^{3x} - \frac{1}{3}\cdot\frac{1}{3}\int e^{3x}3\,dx$ Before taking the integral of $\frac{1}{3}e^{3x}\,dx$, you must put in a 3 and another 1/3.

$\int xe^{3x}\,dx = \frac{1}{3}xe^{3x} - \frac{1}{9}e^{3x} + c$ Integrate to arrive at the solution.

2. $\int \ln 4x\,dx$ Note that no other integration method will work.

$dv = dx \quad u = \ln 4x$ Make the *dv* and *u* substitutions.

$v = x \quad du = \dfrac{1}{4x} \bullet 4 = \dfrac{1}{x} dx$ 　　Integrate dx to get x.

Differentiate ln$4x$ to get $(1/x)\,dx$.

$\int \ln 4x\, dx = \ln 4x \bullet x - \int x \bullet \dfrac{1}{x} dx$

Rewrite the formula above.

$\int \ln 4x\, dx = \ln 4x \bullet x - \int dx$　　Simplify the integral.

$\int \ln 4\, dx = \ln 4x \bullet x - x + c$

Integrate dx to get the value $x + c$

Taking the integral of a function and evaluating it from one x value to another provides the **total area under the curve** (i.e. between the curve and the x axis). Remember, though, that regions above the x axis have "positive" area and regions below the x axis have "negative" area. You must account for these positive and negative values when finding the area under curves. Follow these steps.

1. Determine the x values that will serve as the left and right boundaries of the region.
2. Find all x values between the boundaries that are either solutions to the function or are values which are not in the domain of the function. These numbers are the interval numbers.
3. Integrate the function.
4. Evaluate the integral once for each of the intervals using the boundary numbers.
5. If any of the intervals evaluates to a negative number, make it positive (the negative simply tells you that the region is below the x axis).
6. Add the value of each integral to arrive at the area under the curve.

Example:
Find the area under the following function on the given intervals.

$f(x) = \sin x\,;\ (0, 2\pi)$
$\sin x = 0$　　　　　　　Find any roots to f(x) on $(0, 2\pi)$.
$x = \pi$
$(0, \pi)\ \ (\pi, 2\pi)$　　　　Determine the intervals using the boundary numbers and the roots.

TEACHER CERTIFICATION STUDY GUIDE

$\int \sin x\, dx = -\cos x$

Integrate f(x).

We can ignore the constant c because we have numbers to use to evaluate the integral.

$-\cos x \Big]_{x=0}^{x=\pi} = -\cos \pi - (-\cos 0)$

$-\cos x \Big]_{x=0}^{x=\pi} = -(-1) + (1) = 2$

$-\cos x \Big]_{x=\pi}^{x=2\pi} = -\cos 2\pi - (-\cos \pi)$

$-\cos x \Big]_{x=\pi}^{x=2\pi} = -1 + (-1) = -2$

The -2 means that for $(\pi, 2\pi)$, the region is below the x axis, but the area is still 2. Add the 2 integrals together to get the area.

Area $= 2 + 2 = 4$

Finding the **area between two curves** is much the same as finding the area under one curve. But instead of finding the roots of the functions, you need to find the x values which produce the same number from both functions (set the functions equal and solve). Use these numbers and the given boundaries to write the intervals. On each interval you must pick sample values to determine which function is "on top" of the other. Find the integral of each function. For each interval, subtract the "bottom" integral from the "top" integral. Use the interval numbers to evaluate each of these differences. Add the evaluated integrals to get the total area between the curves.

Example:

Find the area of the regions bounded by the two functions on the indicated intervals.

$f(x) = x + 2$ and $g(x) = x^2$ $[-2, 3]$

Set the functions equal to each other.

$x + 2 = x^2$

$0 = x^2 - x - 2$

$0 = (x - 2)(x + 1)$

$x = 2$ or $x = -1$

Use the solutions and the

MATHEMATICS 130

$(^-2,^-1) \quad (^-1,2) \quad (2,3)$ — boundary numbers to write the intervals.

$$f(^-3/2) = \left(\frac{-3}{2}\right) + 2 = \frac{1}{2}$$

Pick sample values on the integral and evaluate each function as that number.

$$g(^-3/2) = \left(\frac{-3}{2}\right)^2 = \frac{9}{4}$$

$g(x)$ is "on top" on $\left[^-2,^-1\right]$.

$$f(0) = 2$$

$f(x)$ is "on top" on $\left[^-1,2\right]$.

$$g(0) = 0$$

$$f(5/2) = \frac{5}{2} + 2 = \frac{9}{2}$$

$g(x)$ is "on top" on $[2,3]$.

$$g(5/2) = \left(\frac{5}{2}\right)^2 = \frac{25}{4}$$

$$\int f(x)dx = \int (x+2)dx$$

$$\int f(x)dx = \int x dx + 2\int dx$$

$$\int f(x)dx = \frac{1}{1+1}x^{1+1} + 2x$$

$$\int f(x)dx = \frac{1}{2}x^2 + 2x$$

$$\int g(x)dx = \int x^2 dx$$

$$\int g(x)dx = \frac{1}{2+1}x^{2+1} = \frac{1}{3}x^3$$

Area $1 = \int g(x)dx - \int f(x)dx$ $g(x)$ is "on top" on $\left[^-2,^-1\right]$.

Area $1 = \frac{1}{3}x^3 - \left(\frac{1}{2}x^2 + 2x\right)\Big]_{-2}^{-1}$

Area $1 = \left[\frac{1}{3}(^-1)^3 - \left(\frac{1}{2}(^-1)^2 + 2(^-1)\right)\right] - \left[\frac{1}{3}(^-2)^3 - \left(\frac{1}{2}(^-2)^2 + 2(^-2)\right)\right]$

Area $1 = \left[\frac{-1}{3} - \left(\frac{-3}{2}\right)\right] - \left[\frac{-8}{3} - (^-2)\right]$

Area $1 = \left(\frac{7}{6}\right) - \left(\frac{-2}{3}\right) = \frac{11}{6}$

Area 2 = $\int f(x)dx - \int g(x)dx$ $f(x)$ is "on top" on $[^-1,2]$.

Area 2 = $\frac{1}{2}x^2 + 2x - \frac{1}{3}x^3 \Big]_{-1}^{2}$

Area 2 = $\left(\frac{1}{2}(2)^2 + 2(2) - \frac{1}{3}(2)^3\right) - \left(\frac{1}{2}(^-1)^2 + 2(^-1) - \frac{1}{3}(^-1)^3\right)$

Area 2 = $\left(\frac{10}{3}\right) - \left(\frac{1}{2} - 2 + \frac{1}{3}\right)$

Area 2 = $\frac{27}{6}$

Area 3 = $\int g(x)dx - \int f(x)dx$ $g(x)$ is "on top" on $[2,3]$.

Area 3 = $\frac{1}{3}x^3 - \left(\frac{1}{2}x^2 + 2x\right)\Big]_{2}^{3}$

Area 3 = $\left[\frac{1}{3}(3)^3 - \left(\frac{1}{2}(3^2) + 2(3)\right)\right] - \left[\frac{1}{3}(2)^3 - \left(\frac{1}{2}(2)^2 + 2(2)\right)\right]$

Area 3 = $\left(\frac{-3}{2}\right) - \left(\frac{-10}{3}\right) = \frac{11}{6}$

Total area = $\frac{11}{6} + \frac{27}{6} + \frac{11}{6} = \frac{49}{6} = 8\frac{1}{6}$

If you take the area bounded by a curve or curves and revolve it about a line, the result is a solid of revolution. To find the volume of such a solid, the Washer Method works in most instances. Imagine slicing through the solid perpendicular to the line of revolution. The "slice" should resemble a washer. Use an integral and the formula for the volume of disk.

$$Volume_{disk} = \pi \cdot radius^2 \cdot thickness$$

Depending on the situation, the radius is the distance from the line of revolution to the curve; or if there are two curves involved, the radius is the difference between the two functions. The thickness is *dx* if the line of revolution is parallel to the *x* axis and *dy* if the line of revolution is parallel to the *y* axis. Finally, integrate the volume expression using the boundary numbers from the interval.

Example:

Find the value of the solid of revolution found by revolving $f(x) = 9 - x^2$ about the *x* axis on the interval $[0, 4]$.

radius $= 9 - x^2$
thickness $= dx$

$Volume = \int_0^4 \pi(9 - x^2)^2 dx$ Use the formula for volume of a disk.

$Volume = \pi \int_0^4 (81 - 18x^2 + x^4) dx$

$Volume = \pi \left(81x - \dfrac{18}{2+1} x^3 + \dfrac{1}{4+1} x^5 \right) \Big]_0^4$ Take the integral.

$Volume = \pi \left(81x - 6x^3 + \dfrac{1}{5} x^5 \right) \Big]_0^4$

Evaluate the integral first $x = 4$ then at $x = 0$

$Volume = \pi \left[\left(324 - 384 + \dfrac{1024}{5} \right) - (0 - 0 + 0) \right]$

$Volume = \pi \left(144 \dfrac{4}{5} \right) = 144 \dfrac{4}{5} \pi = 454.9$

0007   DATA ANALYSIS, STATISTICS AND PROBABILITY

Percentiles divide data into 100 equal parts. A person whose score falls in the 65th percentile has outperformed 65 percent of all those who took the test. This does not mean that the score was 65 percent out of 100 nor does it mean that 65 percent of the questions answered were correct. It means that the grade was higher than 65 percent of all those who took the test.

Stanine "standard nine" scores combine the understandability of percentages with the properties of the normal curve of probability. Stanines divide the bell curve into nine sections, the largest of which stretches from the 40th to the 60th percentile and is the "Fifth Stanine" (the average of taking into account error possibilities).

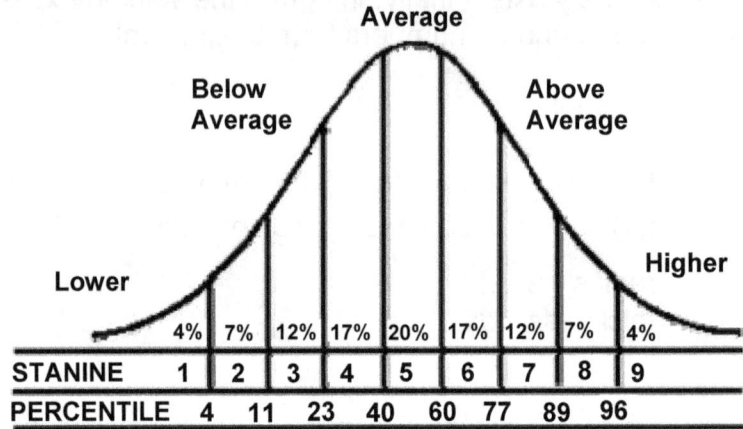

Quartiles divide the data into 4 parts. First find the median of the data set (Q2), then find the median of the upper (Q3) and lower (Q1) halves of the data set. If there are an odd number of values in the data set, include the median value in both halves when finding quartile values. For example, given the data set: {1, 4, 9, 16, 25, 36, 49, 64, 81} first find the median value, which is 25 this is the second quartile. Since there are an odd number of values in the data set (9), we include the median in both halves. To find the quartile values, we much find the medians of: {1, 4, 9, 16, 25} and {25, 36, 49, 64, 81}. Since each of these subsets had an odd number of elements (5), we use the middle value.

Thus the first quartile value is 9 and the third quartile value is 49. If the data set had an even number of elements, average the middle two values. The quartile values are always either one of the data points, or exactly half way between two data points.

Sample problem:

1. Given the following set of data, find the percentile of the score 104.

    70, 72, 82, 83, 84, 87, 100, 104, 108, 109, 110, 115

    Solution: Find the percentage of scores below 104.

    7/12 of the scores are less than 104. This is 58.333%; therefore, the score of 104 is in the 58th percentile.

2. Find the first, second and third quartile for the data listed.

  6, 7, 8, 9, 10, 12, 13, 14, 15, 16, 18, 23, 24, 25, 27, 29, 30, 33, 34, 37

    Quartile 1: The 1st Quartile is the median of the lower half of the data set, which is 11.

    Quartile 2: The median of the data set is the 2nd Quartile, which is 17.

    Quartile 3: The 3rd Quartile is the median of the upper half of the data set, which is 28.

Dependent events occur when the probability of the second event depends on the outcome of the first event. For example, consider the two events (A) it is sunny on Saturday and (B) you go to the beach.

If you intend to go to the beach on Saturday, rain or shine, then A and B may be independent. If however, you plan to go to the beach only if it is sunny, then A and B may be dependent. In this situation, the probability of event B will change depending on the outcome of event A.

Suppose you have a pair of dice, one red and one green. If you roll a three on the red die and then roll a four on the green die, we can see that these events do not depend on the other. The total probability of the two independent events can be found by multiplying the separate probabilities.

$$P(A \text{ and } B) = P(A) \times P(B)$$
$$= 1/6 \times 1/6$$
$$= 1/36$$

Many times, however, events are not independent. Suppose a jar contains 12 red marbles and 8 blue marbles. If you randomly pick a red marble, replace it and then randomly pick again, the probability of picking a red marble the second time remains the same. However, if you pick a red marble, and then pick again without replacing the first red marble, the second pick becomes dependent upon the first pick.

P(Red and Red) with replacement = P(Red) × P(Red)
$$= 12/20 \times 12/20$$
$$= 9/25$$

P(Red and Red) without replacement = P(Red) × P(Red)
$$= 12/20 \times 11/19$$
$$= 33/95$$

Odds are defined as the ratio of the number of favorable outcomes to the number of unfavorable outcomes. The sum of the favorable outcomes and the unfavorable outcomes should always equal the total possible outcomes.

For example, given a bag of 12 red and 7 green marbles compute the odds of randomly selecting a red marble.

$$\text{Odds of red} = \frac{12}{19}$$

$$\text{Odds of not getting red} = \frac{7}{19}$$

In the case of flipping a coin, it is equally likely that a head or a tail will be tossed. The odds of tossing a head are 1:1. This is called even odds.

Mean, median and mode are three measures of central tendency. The **mean** is the average of the data items. The **median** is found by putting the data items in order from smallest to largest and selecting the item in the middle (or the average of the two items in the middle). The **mode** is the most frequently occurring item.
Range is a measure of variability. It is found by subtracting the smallest value from the largest value.

Sample problem:

Find the mean, median, mode and range of the test score listed below:

85	77	65
92	90	54
88	85	70
75	80	69
85	88	60
72	74	95

Mean (X) = sum of all scores ÷ number of scores
        = 78

Median = put numbers in order from smallest to largest. Pick middle number.
54, 60, 65, 69, 70, 72, 74, 75, 77, 80, 85, 85, 85, 88, 88, 90, 92, 95
                                            -- --
                                 both in middle
Therefore, median is average of two numbers in the middle or 78.5

Mode = most frequent number
      = 85

Range = largest number minus the smallest number
       = 95 − 54
       = 41

Different situations require different information. If we examine the circumstances under which an ice cream store owner may use statistics collected in the store, we find different uses for different information.

Over a 7-day period, the store owner collected data on the ice cream flavors sold. He found the mean number of scoops sold was 174 per day. The most frequently sold flavor was vanilla. This information was useful in determining how much ice cream to order in all and in what amounts for each flavor.

In the case of the ice cream store, the median and range had little business value for the owner.

Consider the set of test scores from a math class: 0, 16, 19, 65, 65, 65, 68, 69, 70, 72, 73, 73, 75, 78, 80, 85, 88, and 92. The mean is 64.06 and the median is 71. Since there are only three scores less than the mean out of the eighteen scores, the median (71) would be a more descriptive score.

Retail store owners may be most concerned with the most common dress size so they may order more of that size than any other.

Basic statistical concepts can be applied without computations. For example, inferences can be drawn from a graph or statistical data. A bar graph could display which grade level collected the most money. Student test scores would enable the teacher to determine which units need to be remediated.

To make a bar graph or a pictograph, determine the scale to be used for the graph. Then determine the length of each bar on the graph or determine the number of pictures needed to represent each item of information. Be sure to include an explanation of the scale in the legend.

Example: A class had the following grades:
4 A's, 9 B's, 8 C's, 1 D, 3 F's.
Graph these on a bar graph and a pictograph.

Pictograph

Grade	Number of Students
A	☺☺☺☺
B	☺☺☺☺☺☺☺☺☺
C	☺☺☺☺☺☺☺☺
D	☺
F	☺☺☺

Bar graph

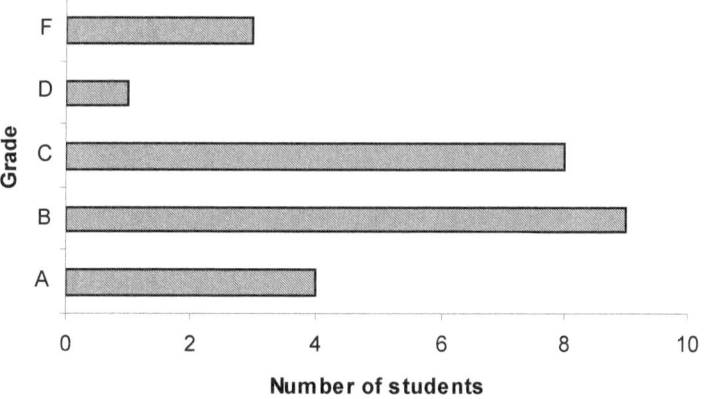

To make a **line graph**, determine appropriate scales for both the vertical and horizontal axes (based on the information to be graphed). Describe what each axis represents and mark the scale periodically on each axis. Graph the individual points of the graph and connect the points on the graph from left to right.

Example: Graph the following information using a line graph.

The number of National Merit finalists/school year

	90-'91	91-'92	92-'93	93-'94	94-'95	95-'96
Central	3	5	1	4	6	8
Wilson	4	2	3	2	3	2

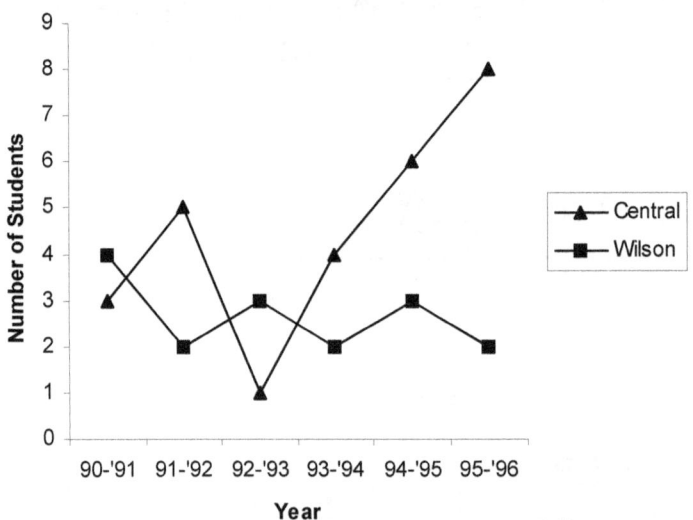

To make a **circle graph**, total all the information that is to be included on the graph. Determine the central angle to be used for each sector of the graph using the following formula:

$$\frac{\text{information}}{\text{total information}} \times 360° = \text{degrees in central} \sphericalangle$$

Lay out the central angles to these sizes, label each section and include its percent.

Example: Graph this information on a circle graph:

Monthly expenses:

Rent, $400
Food, $150
Utilities, $75
Clothes, $75
Church, $100
Misc., $200

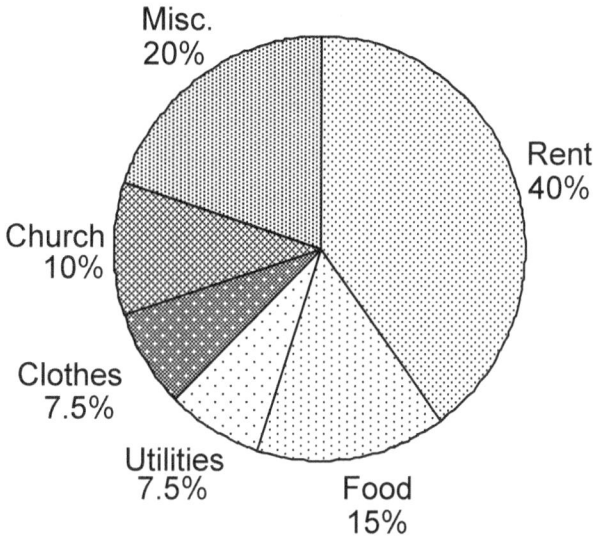

To read a **bar graph or a pictograph**, read the explanation of the scale that was used in the legend. Compare the length of each bar with the dimensions on the axes and calculate the value each bar represents. On a pictograph count the number of pictures used in the chart and calculate the value of all the pictures.

To read a circle graph, find the total of the amounts represented on the entire circle graph. To determine the actual amount that each sector of the graph represents, multiply the percent in a sector times the total amount number.

To read a chart read the row and column headings on the table. Use this information to evaluate the given information in the chart.

0008   **MATRIX ALGEBRA**

Word problems can sometimes be solved by using a system of two equations in 2 unknowns. This system can then be solved using **substitution**, the **addition-subtraction method**, or **determinants**.

Example: Farmer Greenjeans bought 4 cows and 6 sheep for $1700. Mr. Ziffel bought 3 cows and 12 sheep for $2400. If all the cows were the same price and all the sheep were another price, find the price charged for a cow or for a sheep.

Let x = price of a cow
Let y = price of a sheep

Then Farmer Greenjeans' equation would be: $4x + 6y = 1700$
Mr. Ziffel's equation would be: $3x + 12y = 2400$

To solve by **addition-subtraction**:
Multiply the first equation by $^-2$: $^-2(4x + 6y = 1700)$
Keep the other equation the same: $(3x + 12y = 2400)$
By doing this, the equations can be added to each other to eliminate one variable and solve for the other variable.

$$^-8x - 12y = ^-3400$$
$$3x + 12y = 2400 \quad \text{Add these equations.}$$
$$^-5x = ^-1000$$

$x = 200 \leftarrow$ the price of a cow was $200.
Solving for y, $y = 150 \leftarrow$ the price of a sheep, $150.

To solve by **substitution**:
Solve one of the equations for a variable. (Try to make an equation without fractions if possible.) Substitute this expression into the equation that you have not yet used. Solve the resulting equation for the value of the remaining variable.

$$4x + 6y = 1700$$
$$3x + 12y = 2400 \leftarrow \text{Solve this equation for } x.$$

It becomes $x = 800 - 4y$. Now substitute $800 - 4y$ in place of x in the OTHER equation. $4x + 6y = 1700$ now becomes:
$$4(800 - 4y) + 6y = 1700$$
$$3200 - 16y + 6y = 1700$$
$$3200 - 10y = 1700$$
$$^-10y = {}^-1500$$
$$y = 150, \text{ or } \$150 \text{ for a sheep.}$$

Substituting 150 back into an equation for y, find x.
$$4x + 6(150) = 1700$$
$$4x + 900 = 1700$$
$$4x = 800 \text{ so } x = 200 \text{ for a cow.}$$

To solve by **determinants**:

Let x = price of a cow
Let y = price of a sheep

Then Farmer Greenjeans' equation would be: $4x + 6y = 1700$
Mr. Ziffel's equation would be: $3x + 12y = 2400$

To solve this system using determinants, make one 2 by 2 determinant divided by another 2 by 2 determinant. The bottom determinant is filled with the x and y term coefficients. The top determinant is almost the same as this bottom determinant. The only difference is that when you are solving for x, the x coefficients are replaced with the constants. Likewise, when you are solving for y, the y coefficients are replaced with the constants. To find the value of a 2 by 2 determinant, $\begin{pmatrix} a & b \\ c & d \end{pmatrix}$, is found by $ad - bc$.

$$x = \frac{\begin{pmatrix} 1700 & 6 \\ 2400 & 12 \end{pmatrix}}{\begin{pmatrix} 4 & 6 \\ 3 & 12 \end{pmatrix}} = \frac{1700(12) - 6(2400)}{4(12) - 6(3)} = \frac{20400 - 14400}{48 - 18} = \frac{6000}{30} = 200$$

$$y = \frac{\begin{pmatrix} 4 & 1700 \\ 3 & 2400 \end{pmatrix}}{\begin{pmatrix} 4 & 6 \\ 3 & 12 \end{pmatrix}} = \frac{2400(4) - 3(1700)}{4(12) - 6(3)} = \frac{9600 - 5100}{48 - 18} = \frac{4500}{30} = 150$$

NOTE: The bottom determinant is always the same value for each letter.

Word problems can sometimes be solved by using a system of three equations in 3 unknowns. This system can then be solved using **substitution**, the **addition-subtraction method**, or **determinants**.

To solve by **substitution**:

Example: Mrs. Allison bought 1 pound of potato chips, a 2 pound beef roast, and 3 pounds of apples for a total of $ 8.19. Mr. Bromberg bought a 3 pound beef roast and 2 pounds of apples for $ 9.05. Kathleen Kaufman bought 2 pounds of potato chips, a 3 pound beef roast, and 5 pounds of apples for $ 13.25. Find the per pound price of each item.

Let x = price of a pound of potato chips
Let y = price of a pound of roast beef
Let z = price of a pound of apples

Mrs. Allison's equation would be: $1x + 2y + 3z = 8.19$
Mr. Bromberg's equation would be: $3y + 2z = 9.05$
K. Kaufman's equation would be: $2x + 3y + 5z = 13.25$

Take the first equation and solve it for x. (This was chosen because x is the easiest variable to get alone in this set of equations.) This equation would become:

$$x = 8.19 - 2y - 3z$$

Substitute this expression into the other equations in place of the letter x:

$$3y + 2z = 9.05 \leftarrow \text{equation 2}$$
$$2(8.19 - 2y - 3z) + 3y + 5z = 13.25 \leftarrow \text{equation 3}$$

Simplify the equation by combining like terms:

$$3y + 2z = 9.05 \leftarrow \text{equation 2}$$
$$*\ {}^-1y - 1z = {}^-3.13 \leftarrow \text{equation 3}$$

Solve equation 3 for either y or z:

$y = 3.13 - z$ Substitute this into equation 2 for y:

$$3(3.13 - z) + 2z = 9.05 \leftarrow \text{equation 2}$$
$${}^-1y - 1z = {}^-3.13 \leftarrow \text{equation 3}$$

Combine like terms in equation 2:

$9.39 - 3z + 2z = 9.05$
$z = .34$ per pound price of apples

Substitute .34 for z in the starred equation above to solve for y:

$y = 3.13 - z$ becomes $y = 3.13 - .34$, so
$y = 2.79 =$ per pound price of roast beef

Substituting .34 for z and 2.79 for y in one of the original equations, solve for x:

$1x + 2y + 3z = 8.19$
$1x + 2(2.79) + 3(.34) = 8.19$
$x + 5.58 + 1.02 = 8.19$
$x + 6.60 = 8.19$
$x = 1.59$ per pound of potato chips

$(x, y, z) = (1.59, 2.79, .34)$

To solve by **addition-subtraction**:

Choose a letter to eliminate. Since the second equation is already missing an x, let's eliminate x from equations 1 and 3.

1) $1x + 2y + 3x = 8.19$ ← Multiply by $^-2$ below.
2) $3y + 2z = 9.05$
3) $2x + 3y + 5z = 13.25$

$^-2(1x + 2y + 3z = 8.19)$ = $^-2x - 4y - 6z = ^-16.38$
Keep equation 3 the same : $2x + 3y + 5z = 13.25$

By doing this, the equations $^-y - z = ^-3.13$ ← equation 4
can be added to each other to
eliminate one variable.

The equations left to solve are equations 2 and 4:
   $^-y - z = ^-3.13$ ← equation 4
   $3y + 2z = 9.05$ ← equation 2

Multiply equation 4 by 3: $3(^-y - z = ^-3.13)$
Keep equation 2 the same: $3y + 2z = 9.05$

$^-3y - 3z = ^-9.39$
$3y + 2z = 9.05$ Add these equations.
$^-1z = ^-.34$
     $z = .34$ ← the per pound price of apples
solving for y, $y = 2.79$ ← the per pound roast beef price
solving for x, $x = 1.59$ ← potato chips, per pound price

To solve by **substitution**:

Solve one of the 3 equations for a variable. (Try to make an equation without fractions if possible.) Substitute this expression into the other 2 equations that you have not yet used.

1) $1x + 2y + 3z = 8.19$ ← Solve for x.
2) $3y + 2z = 9.05$
3) $2x + 3y + 5z = 13.25$
   Equation 1 becomes $x = 8.19 - 2y - 3z$.

Substituting this into equations 2 and 3, they become:

2) $3y + 2z = 9.05$
3) $2(8.19 - 2y - 3z) + 3y + 5z = 13.25$
       $16.38 - 4y - 6z + 3y + 5z = 13.25$
       $^-y - z = {^-}3.13$

The equations left to solve are :

$3y + 2z = 9.05$

$^-y - z = {^-}3.13$ ← Solve for either y or z.

It becomes $y = 3.13 - z$. Now substitute $3.13 - z$ in place of y in the OTHER equation. $3y + 2z = 9.05$ now becomes:

$$3(3.13 - z) + 2z = 9.05$$
$$9.39 - 3z + 2z = 9.05$$
$$9.39 - z = 9.05$$
$$^-z = {^-}.34$$
$$z = .34 \text{, or } \$.34/\text{lb of apples}$$

Substituting .34 back into an equation for z, find y.
     $3y + 2z = 9.05$
     $3y + 2(.34) = 9.05$
     $3y + .68 = 9.05$ so $y = 2.79/\text{lb of roast beef}$

Substituting .34 for z and 2.79 for y into one of the original equations, it becomes:

     $2x + 3y + 5z = 13.25$
     $2x + 3(2.79) + 5(.34) = 13.25$
     $2x + 8.37 + 1.70 = 13.25$
     $2x + 10.07 = 13.25$, so $x = 1.59/\text{lb of potato chips}$

To solve by **determinants**:

Let x = price of a pound of potato chips
Let y = price of a pound of roast beef
Let z = price of a pound of apples
1) $1x + 2y + 3z = 8.19$
2) $3y + 2z = 9.05$
3) $2x + 3y + 5z = 13.25$

To solve this system using determinants, make one 3 by 3 determinant divided by another 3 by 3 determinant. The bottom determinant is filled with the x, y, and z term coefficients. The top determinant is almost the same as this bottom determinant. The only difference is that when you are solving for x, the x coefficients are replaced with the constants. When you are solving for y, the y coefficients are replaced with the constants. Likewise, when you are solving for z, the z coefficients are replaced with the constants. To find the value of a 3 by 3 determinant,

$\begin{pmatrix} a & b & c \\ d & e & f \\ g & h & i \end{pmatrix}$ is found by the following steps:

Copy the first two columns to the right of the determinant:

$\begin{pmatrix} a & b & c \\ d & e & f \\ g & h & i \end{pmatrix} \begin{matrix} a & b \\ d & e \\ g & h \end{matrix}$

Multiply the diagonals from top left to bottom right, and add these diagonals together.

$\begin{pmatrix} a^* & b^\circ & c^\bullet \\ d & e^* & f^\circ \\ g & h & i^* \end{pmatrix} \begin{matrix} a & b \\ d^\bullet & e \\ g^\circ & h^\bullet \end{matrix} = a^* e^* i^* + b^\circ f^\circ g^\circ + c^\bullet d^\bullet h^\bullet$

Then multiply the diagonals from bottom left to top right, and add these diagonals together.

MATHEMATICS

TEACHER CERTIFICATION STUDY GUIDE

$$\begin{pmatrix} a & b & c^* \\ d & e^* & f^\circ \\ g^* & h^\circ & i^\bullet \end{pmatrix} \begin{matrix} a^\circ & b^\bullet \\ d^\bullet & e \\ g & h \end{matrix} = g^* e^* c^* + h^\circ f^\circ a^\circ + i^\bullet d^\bullet b^\bullet$$

Subtract the first diagonal total minus the second diagonal total:

$$(= a^* e^* i^* + b^\circ f^\circ g^\circ + c^\bullet d^\bullet h^\bullet) - (= g^* e^* c^* + h^\circ f^\circ a^\circ + i^\bullet d^\bullet b^\bullet)$$

This gives the value of the determinant. To find the value of a variable, divide the value of the top determinant by the value of the bottom determinant.

1) $1x + 2y + 3z = 8.19$
2) $3y + 2z = 9.05$
3) $2x + 3y + 5z = 13.25$

$$x = \frac{\begin{pmatrix} 8.19 & 2 & 3 \\ 9.05 & 3 & 2 \\ 13.25 & 3 & 5 \end{pmatrix}}{\begin{pmatrix} 1 & 2 & 3 \\ 0 & 3 & 2 \\ 2 & 3 & 5 \end{pmatrix}}$$ solve each determinant using the method shown below

Multiply the diagonals from top left to bottom right, and add these diagonals together.

$$\begin{pmatrix} 8.19^* & 2^\circ & 3^\bullet \\ 9.05 & 3^* & 2^\circ \\ 13.25 & 3 & 5^* \end{pmatrix} \begin{matrix} 8.19 & 2 \\ 9.05^\bullet & 3 \\ 13.25^\circ & 3^\bullet \end{matrix}$$

$$= (8.19^*)(3^*)(5^*) + (2^\circ)(2^\circ)(13.25^\circ) + (3^\bullet)(9.05^\bullet)(3^\bullet)$$

Then multiply the diagonals from bottom left to top right, and add these diagonals together.

MATHEMATICS

$$\begin{pmatrix} 8.19 & 2 & 3^* \\ 9.05 & 3^* & 2^\circ \\ 13.25^* & 3^\circ & 5^\bullet \end{pmatrix} \begin{matrix} 8.19^\circ & 2^\bullet \\ 9.05^\bullet & 3 \\ 13.25 & 3 \end{matrix}$$

$$= (13.25^*)(3^*)(3^*) + (3^\circ)(2^\circ)(8.19^\circ) + (5^\bullet)(9.05^\bullet)(2^\bullet)$$

Subtract the first diagonal total minus the second diagonal total:

$$(8.19^*)(3^*)(5^*) + (2^\circ)(2^\circ)(13.25^\circ) + (3^\bullet)(9.05^\bullet)(3^\bullet) = 257.30$$
$$- (13.25^*)(3^*)(3^*) + (3^\circ)(2^\circ)(8.19^\circ) + (5^\bullet)(9.05^\bullet)(2^\bullet) = {}^-258.89$$
$$\overline{\,{}^-1.59}$$

Use the same multiplying and subtraction procedure for the bottom determinant to get $^-1$ as an answer. Now divide:

$$\frac{^-1.59}{^-1} = \$1.59/\text{lb of potato chips}$$

$$y = \frac{\begin{pmatrix} 1 & 8.19 & 3 \\ 0 & 9.05 & 2 \\ 2 & 13.25 & 5 \end{pmatrix}}{\begin{pmatrix} 1 & 2 & 3 \\ 0 & 3 & 2 \\ 2 & 3 & 5 \end{pmatrix}} = \frac{^-2.79}{^-1} = \$2.79/\text{lb of roast beef}$$

NOTE: The bottom determinant is always the same value for each letter.

$$z = \frac{\begin{pmatrix} 1 & 2 & 8.19 \\ 0 & 3 & 9.05 \\ 2 & 3 & 13.25 \end{pmatrix}}{\begin{pmatrix} 1 & 2 & 3 \\ 0 & 3 & 2 \\ 2 & 3 & 5 \end{pmatrix}} = \frac{^-.34}{^-1} = \$.34/\text{lb of apples}$$

A matrix is a **square array of numbers** called its entries or elements. The dimensions of a matrix are written as the number of rows (r) by the number of columns (r × c).

$$\begin{pmatrix} 1 & 2 & 3 \\ 4 & 5 & 6 \end{pmatrix}$$ is a 2 × 3 matrix (2 rows by 3 columns)

$$\begin{pmatrix} 1 & 2 \\ 3 & 4 \\ 5 & 6 \end{pmatrix}$$ is a 3 × 2 matrix (3 rows by 2 columns)

Associated with every square matrix is a number called the determinant. Use these formulas to calculate determinants.

2 × 2 $\begin{pmatrix} a & b \\ c & d \end{pmatrix} = ad - bc$

3 × 3
$$\begin{pmatrix} a_1 & b_1 & c_1 \\ a_2 & b_2 & c_2 \\ a_3 & b_3 & c_3 \end{pmatrix} = (a_1 b_2 c_3 + b_1 c_2 a_3 + c_1 a_2 b_3) - (a_3 b_2 c_1 + b_3 c_2 a_1 + c_3 a_2 b_1)$$

This is found by repeating the first two columns and then using the diagonal lines to find the value of each expression as shown below:

$$\begin{pmatrix} a_1^* & b_1^\circ & c_1^\bullet \\ a_2 & b_2^* & c_2^\circ \\ a_3 & b_3 & c_3^* \end{pmatrix} \begin{matrix} a_1 & b_1 \\ a_2^\bullet & b_2 \\ a_3^\circ & b_3^\bullet \end{matrix} = (a_1 b_2 c_3 + b_1 c_2 a_3 + c_1 a_2 b_3) - (a_3 b_2 c_1 + b_3 c_2 a_1 + c_3 a_2 b_1)$$

Sample Problem:

1. Find the value of the determinant:

$\begin{pmatrix} 4 & ^-8 \\ 7 & 3 \end{pmatrix} = (4)(3) - (7)(^-8)$ Cross multiply and subtract.

$12 - (^-56) = 68$ Then simplify.

MATHEMATICS

Addition of matrices is accomplished by adding the corresponding elements of the two matrices. Subtraction is defined as the inverse of addition. In other words, change the sign on all the elements in the second matrix and add the two matrices.

Sample problems:

Find the sum or difference.

1. $\begin{pmatrix} 2 & 3 \\ {}^-4 & 7 \\ 8 & {}^-1 \end{pmatrix} + \begin{pmatrix} 8 & {}^-1 \\ 2 & {}^-1 \\ 3 & {}^-2 \end{pmatrix} =$

$\begin{pmatrix} 2+8 & 3+({}^-1) \\ {}^-4+2 & 7+({}^-1) \\ 8+3 & {}^-1+({}^-2) \end{pmatrix}$ Add corresponding elements.

$\begin{pmatrix} 10 & 2 \\ {}^-2 & 6 \\ 11 & {}^-3 \end{pmatrix}$ Simplify.

2. $\begin{pmatrix} 8 & {}^-1 \\ 7 & 4 \end{pmatrix} - \begin{pmatrix} 3 & 6 \\ {}^-5 & 1 \end{pmatrix} =$

$\begin{pmatrix} 8 & {}^-1 \\ 7 & 4 \end{pmatrix} + \begin{pmatrix} {}^-3 & {}^-6 \\ 5 & {}^-1 \end{pmatrix} =$ Change all of the signs in the second matrix and then add the two matrices.

$\begin{pmatrix} 8+({}^-3) & {}^-1+({}^-6) \\ 7+5 & 4+({}^-1) \end{pmatrix} =$ Simplify.

$\begin{pmatrix} 5 & {}^-7 \\ 12 & 3 \end{pmatrix}$

Practice problems:

1. $\begin{pmatrix} 8 & -1 \\ 5 & 3 \end{pmatrix} + \begin{pmatrix} 3 & 8 \\ 6 & -2 \end{pmatrix} =$

2. $\begin{pmatrix} 3 & 7 \\ -4 & 12 \\ 0 & -5 \end{pmatrix} - \begin{pmatrix} 3 & 4 \\ 6 & -1 \\ -5 & -5 \end{pmatrix} =$

Scalar multiplication is the product of the scalar (the outside number) and each element inside the matrix.

Sample problem:

Given: $A = \begin{pmatrix} 4 & 0 \\ 3 & -1 \end{pmatrix}$ Find 2A.

$2A = 2\begin{pmatrix} 4 & 0 \\ 3 & -1 \end{pmatrix}$

$\begin{pmatrix} 2 \times 4 & 2 \times 0 \\ 2 \times 3 & 2 \times -1 \end{pmatrix}$ Multiply each element in the matrix by the scalar.

$\begin{pmatrix} 8 & 0 \\ 6 & -2 \end{pmatrix}$ Simplify.

Practice problems:

3. $-2 \begin{pmatrix} 2 & 0 & 1 \\ -1 & -2 & 4 \end{pmatrix}$

4. $3 \begin{pmatrix} 6 \\ 2 \\ 8 \end{pmatrix} + 4 \begin{pmatrix} 0 \\ 7 \\ 2 \end{pmatrix}$

5. $2 \begin{pmatrix} -6 & 8 \\ -2 & -1 \\ 0 & 3 \end{pmatrix}$

The variable in a matrix equation represents a matrix. When solving for the answer use the adding, subtracting and scalar multiplication properties.

Sample problem:

Solve the matrix equation for the variable X.

$$2x + \begin{pmatrix} 4 & 8 & 2 \\ 7 & 3 & 4 \end{pmatrix} = 2\begin{pmatrix} 1 & ^-2 & 0 \\ 3 & ^-5 & 7 \end{pmatrix}$$

$$2x = 2\begin{pmatrix} 1 & ^-2 & 0 \\ 3 & ^-5 & 7 \end{pmatrix} - \begin{pmatrix} 4 & 8 & 2 \\ 7 & 3 & 4 \end{pmatrix}$$ Subtract $\begin{pmatrix} 4 & 8 & 2 \\ 7 & 3 & 4 \end{pmatrix}$ from both sides.

$$2x = \begin{pmatrix} 2 & ^-4 & 0 \\ 6 & ^-10 & 14 \end{pmatrix} + \begin{pmatrix} ^-4 & ^-8 & ^-2 \\ ^-7 & ^-3 & ^-4 \end{pmatrix}$$ Scalar multiplication and matrix subtraction.

$$2x = \begin{pmatrix} ^-2 & ^-12 & ^-2 \\ ^-1 & ^-13 & 10 \end{pmatrix}$$ Matrix addition.

$$x = \begin{pmatrix} ^-1 & ^-6 & ^-1 \\ -\frac{1}{2} & -\frac{13}{2} & 5 \end{pmatrix}$$ Multiply both sides by .

Solve for the unknown values of the elements in the matrix.

$$\begin{pmatrix} x+3 & y-2 \\ z+3 & w-4 \end{pmatrix} + \begin{pmatrix} ^-2 & 4 \\ 2 & 5 \end{pmatrix} = \begin{pmatrix} 4 & 8 \\ 6 & 1 \end{pmatrix}$$

$$\begin{pmatrix} x+1 & y+2 \\ z+5 & w+1 \end{pmatrix} = \begin{pmatrix} 4 & 8 \\ 6 & 1 \end{pmatrix}$$ Matrix addition.

$x+1=4 \quad y+2=8 \quad z+5=6 \quad w+1=1$
$x=3 \quad\quad y=6 \quad\quad z=1 \quad\quad w=0$ Definition of equal matrices.

MATHEMATICS

Practice problems:

1. $x + \begin{pmatrix} 7 & 8 \\ 3 & -1 \\ 2 & -3 \end{pmatrix} = \begin{pmatrix} 0 & 8 \\ -9 & -4 \\ 8 & 2 \end{pmatrix}$

2. $4x - 2\begin{pmatrix} 0 & 10 \\ 6 & -4 \end{pmatrix} = 3\begin{pmatrix} 4 & 9 \\ 0 & 12 \end{pmatrix}$

3. $\begin{pmatrix} 7 & 3 \\ 2 & 4 \\ 3 & 7 \end{pmatrix} + \begin{pmatrix} a+2 & b+4 \\ c-3 & d+1 \\ e & f+3 \end{pmatrix} = \begin{pmatrix} 4 & 6 \\ -1 & 1 \\ 3 & 0 \end{pmatrix}$

The **product of two matrices** can only be found if the number of columns in the first matrix is equal to the number of rows in the second matrix. Matrix multiplication is not necessarily commutative.

Sample problems:

1. Find the product AB if:

$A = \begin{pmatrix} 2 & 3 & 0 \\ 1 & -4 & -2 \\ 0 & 1 & 1 \end{pmatrix}$ $B = \begin{pmatrix} -2 & 3 \\ 6 & -1 \\ 0 & 2 \end{pmatrix}$

3×3 3×2

Note: Since the number of columns in the first matrix ($3 \times \underline{3}$) matches the number of rows ($\underline{3} \times 2$) this product is defined and can be found. The dimensions of the product will be equal to the number of rows in the first matrix ($\underline{3} \times 3$) by the number of columns in the second matrix ($3 \times \underline{2}$). The answer will be a 3×2 matrix.

$AB = \begin{pmatrix} 2 & 3 & 0 \\ 1 & -4 & -2 \\ 0 & 1 & 1 \end{pmatrix} \times \begin{pmatrix} -2 & 3 \\ 6 & -1 \\ 0 & 2 \end{pmatrix}$

MATHEMATICS

$$\begin{pmatrix} 2(^-2)+3(6)+0(0) & & \\ & & \\ & & \end{pmatrix}$$ Multiply 1ˢᵗ row of A by 1ˢᵗ column of B.

$$\begin{pmatrix} 14 & 2(3)+3(^-1)+0(2) \\ & \\ & \end{pmatrix}$$ Multiply 1ˢᵗ row of A by 2ⁿᵈ column of B.

$$\begin{pmatrix} 14 & 3 \\ 1(^-2)-4(6)-2(0) & \\ & \end{pmatrix}$$ Multiply 2ⁿᵈ row of A by 1ˢᵗ column of B.

$$\begin{pmatrix} 14 & 3 \\ ^-26 & 1(3)-4(^-1)-2(2) \\ & \end{pmatrix}$$ Multiply 2ⁿᵈ row of A by 2ⁿᵈ column of B.

$$\begin{pmatrix} 14 & 3 \\ ^-26 & 3 \\ 0(^-2)+1(6)+1(0) & \end{pmatrix}$$ Multiply 3ʳᵈ row of A by 1ˢᵗ column of B.

$$\begin{pmatrix} 14 & 3 \\ ^-26 & 3 \\ 6 & 0(3)+1(^-1)+1(2) \end{pmatrix}$$ Multiply 3ʳᵈ row of A by 2ⁿᵈ column of B.

$$\begin{pmatrix} 14 & 3 \\ ^-26 & 3 \\ 6 & 1 \end{pmatrix}$$

The product of BA is not defined since the number of columns in B is not equal to the number of rows in A.

Practice problems:

1. $\begin{pmatrix} 3 & 4 \\ ^-2 & 1 \end{pmatrix} \begin{pmatrix} ^-1 & 7 \\ ^-3 & 1 \end{pmatrix}$

2. $\begin{pmatrix} 1 & ^-2 \\ 3 & 4 \\ 2 & 5 \\ -1 & 6 \end{pmatrix} \begin{pmatrix} 3 & ^-1 & ^-4 \\ ^-1 & 2 & 3 \end{pmatrix}$

When given the following system of equations:

$$ax + by = e$$
$$cx + dy = f$$

the matrix equation is written in the form:

$$\begin{pmatrix} a & b \\ c & d \end{pmatrix} \begin{pmatrix} x \\ y \end{pmatrix} = \begin{pmatrix} e \\ f \end{pmatrix}$$

The solution is found using the inverse of the matrix of coefficients. Inverse of matrices can be written as follows:

$$A^{-1} = \frac{1}{\text{determinant of } A} \begin{pmatrix} d & -b \\ -c & a \end{pmatrix}$$

Sample Problem:
1. Write the matrix equation of the system.

$$3x - 4y = 2$$
$$2x + y = 5$$

$$\begin{pmatrix} 3 & -4 \\ 2 & 1 \end{pmatrix} \begin{pmatrix} x \\ y \end{pmatrix} = \begin{pmatrix} 2 \\ 5 \end{pmatrix}$$ Definition of matrix equation.

$$\begin{pmatrix} x \\ y \end{pmatrix} = \frac{1}{11} \begin{pmatrix} 1 & 4 \\ -2 & 3 \end{pmatrix} \begin{pmatrix} 2 \\ 5 \end{pmatrix}$$ Multiply by the inverse of the coefficient matrix.

$$\begin{pmatrix} x \\ y \end{pmatrix} = \frac{1}{11} \begin{pmatrix} 22 \\ 11 \end{pmatrix}$$ Matrix multiplication.

$$\begin{pmatrix} x \\ y \end{pmatrix} = \begin{pmatrix} 2 \\ 1 \end{pmatrix}$$ Scalar multiplication.

The solution is (2,1).

Practice problems:

1. $x + 2y = 5$
   $3x + 5y = 14$

2. $-3x + 4y - z = 3$
   $x + 2y - 3z = 9$
   $y - 5z = -1$

0009   DISCRETE MATHEMATICS

When given a set of numbers where the common difference between the terms is constant, use the following formula:

$$a_n = a_1 + (n-1)d$$ where a_1 = the first term
n = the n th term (general term)
d = the common difference

Sample problem:

1. Find the 8th term of the arithmetic sequence 5, 8, 11, 14, ...

    $a_n = a_1 + (n-1)d$
    $a_1 = 5$ Identify 1st term.
    $d = 3$ Find d.
    $a_8 = 5 + (8-1)3$ Substitute.
    $a_8 = 26$

2. Given two terms of an arithmetic sequence find a and d.

    $a_4 = 21$ $a_6 = 32$
    $a_n = a_1 + (n-1)d$
    $21 = a_1 + (4-1)d$
    $32 = a_1 + (6-1)d$

    $21 = a_1 + 3d$ Solve the system of equations.
    $32 = a_1 + 5d$

    $21 = a_1 + 3d$
    $-32 = {}^-a_1 - 5d$ Multiply by $^-1$ and add the equations.
    $\overline{{}^-11 = {}^-2d}$
    $5.5 = d$

    $21 = a_1 + 3(5.5)$ Substitute $d = 5.5$ into one of the equations.
    $21 = a_1 + 16.5$
    $a_1 = 4.5$

The sequence begins with 4.5 and has a common difference of 5.5 between numbers.

When using geometric sequences consecutive numbers are compared to find the common ratio.

$$r = \frac{a_{n+1}}{a_n}$$

$r = $ the common ratio
$a_n = $ the nth term

The ratio is then used in the geometric sequence formula:

$$a_n = a_1 r^{n-1}$$

Sample problems:

1. Find the 8th term of the geometric sequence 2, 8, 32, 128 ...

$r = \dfrac{a_{n+1}}{a_n}$ Use the common ratio formula to find r.

$r = \dfrac{8}{2} = 4$ Substitute $a_n = 2$ $a_{n+1} = 8$

$a_n = a_1 \times r^{n-1}$ Use $r = 4$ to solve for the 8th term.
$a_8 = 2 \times 4^{8-1}$
$a_8 = 32768$

The sums of terms in a progression is simply found by determining if it is an arithmetic or geometric sequence and then using the appropriate formula.

Sum of first n terms of an arithmetic sequence.

$$S_n = \frac{n}{2}(a_1 + a_n)$$

or

$$S_n = \frac{n}{2}\left[2a_1 + (n-1)d\right]$$

Sum of first n terms of a geometric sequence.

$$S_n = \frac{a_1(r^n - 1)}{r - 1}, r \neq 1$$

Sample Problems:

1. $\sum_{i=1}^{10}(2i + 2)$

   This means find the sum of the terms beginning with the first term and ending with the 10th term of the sequence $a = 2i + 2$.

   $a_1 = 2(1) + 2 = 4$
   $a_{10} = 2(10) + 2 = 22$

   $S_n = \frac{n}{2}(a_1 + a_n)$
   $S_n = \frac{10}{2}(4 + 22)$
   $S_n = 130$

2. Find the sum of the first 6 terms in an arithmetic sequence if the first term is 2 and the common difference, d, is -3.

   $n = 6 \qquad a_1 = 2 \qquad d = {}^-3$

   $S_n = \frac{n}{2}\left[2a_1 + (n-1)d\right]$

   $S_6 = \frac{6}{2}\left[2 \times 2 + (6-1){}^-3\right]$ Substitute known values.

   $S_6 = 3\left[4 + ({}^-15)\right]$ Solve.

   $S_6 = 3(-11) = -33$

MATHEMATICS

3. Find $\sum_{i=1}^{5} 4 \times 2^i$

This means the sum of the first 5 terms where $a_i = a \times b^i$ and $r = b$.

$a_1 = 4 \times 2^1 = 8$ Identify a_1, r, n
$r = 2 \quad n = 5$

$S_n = \dfrac{a_1(r^n - 1)}{r - 1}$ Substitute a, r, n

$S_5 = \dfrac{8(2^5 - 1)}{2 - 1}$ Solve.

$S_5 = \dfrac{8(31)}{1} = 248$

Practice problems:

1. Find the sum of the first five terms of the sequence if $a = 7$ and $d = 4$.

2. $\sum_{i=1}^{7} (2i - 4)$

3. $\sum_{i=1}^{6} -3\left(\dfrac{2}{5}\right)^i$

The difference between permutations and combinations is that in permutations all possible ways of writing an arrangement of objects are given while in a combination a given arrangement of objects is listed only once.

Given the set {1, 2, 3, 4}, list the arrangements of two numbers that can be written as a combination and as a permutation.

Combination	Permutation
12, 13, 14, 23, 24, 34	12, 21, 13, 31, 14, 41, 23, 32, 24, 42, 34, 43,
six ways	twelve ways

Using the formulas given below the same results can be found.

$$_nP_r = \frac{n!}{(n-r)!}$$

The notation $_nP_r$ is read "the number of permutations of n objects taken r at a time."

$$_4P_2 = \frac{4!}{(4-2)!}$$

Substitute known values.

$$_4P_2 = 12$$

Solve.

$$_nC_r = \frac{n!}{(n-r)!r!}$$

The number of combinations when r objects are selected from n objects.

$$_4C_2 = \frac{4!}{(4-2)!2!}$$

Substitute known values.

$$_4C_2 = 6$$

Solve.

An **arithmetic sequence** is a sequence where each successive term is obtained from the previous term by addition or subtraction of a fixed number called a **difference**. In order for it to be an arithmetic sequence the SAME number must be added, or subtracted, if that is the pattern, each time. An example would be 1, 5, 9, 13, ... where 4 is added to each previous number.

A **geometric sequence** is a sequence where each successive term is obtained from the previous term by multiplying by a fixed number called a **ratio**. In order for it to be a geometric sequence, each term must be multiplied by the SAME number. An example would be 2, 8, 32, 128, ... where each term is multiplied by 4.

Possibly the most famous sequence is the **Fibonacci sequence**. A basic Fibonacci sequence is when two numbers are added together to get the next number in the sequence. An example would be 1, 1, 2, 3, 5, 8, 13, ...

Sequences can be **finite** or **infinite**. A finite sequence is a sequence whose domain consists of the set {1, 2, 3, ... n} or the first *n* positive integers. An infinite sequence is a sequence whose domain consists of the set {1, 2, 3, ...}; which is in other words all positive integers.

A **recurrence relation** is an equation that defines a sequence recursively; in other words, each term of the sequence is defined as a function of the preceding terms.

A real-life application would be using a recurrence relation to determine how much your savings would be in an account at the end of a certain period of time. For example:

You deposit $5,000 in your savings account. Your bank pays 5% interest compounded annually. How much will your account be worth at the end of 10 years?

Let V represent the amount of money in the account and V_n represent the amount of money after *n* years.

The amount in the account after *n* years equals the amount in the account after $n - 1$ years plus the interest for the *n*th year. This can be expressed as the recurrence relation V_0 where your initial deposit is represented by $V_0 = 5,000$.

$$V_0 = V_0$$
$$V_1 = 1.05 V_0$$
$$V_2 = 1.05 V_1 = (1.05)^2 V_0$$
$$V_3 = 1.05 V_2 = (1.05)^3 V_0$$
......
$$V_n = (1.05) V_{n-1} = (1.05)^n V_0$$

Inserting the values into the equation, you get
$$V_{10} = (1.05)^{10}(5,000) = 8,144.$$

You determine that after investing $5,000 in an account earning 5% interest, compounded annually for 10 years, you would have $8,144.

A **pictograph** uses small figures or icons to represent data. Pictographs are used to summarize relative amounts, trends, and data sets. They are useful in comparing quantities.

Monarch Butterfly Migration to the U.S.
in millions

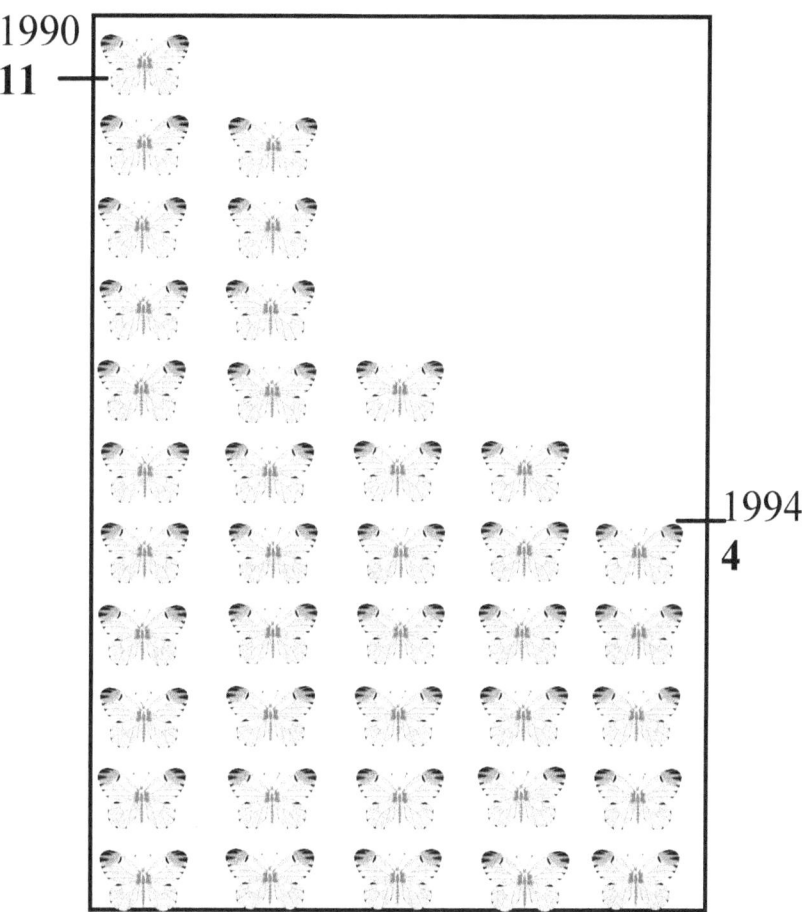

The data in this graph is not accurate. It is for illustration purposes only.

Scatter plots compare two characteristics of the same group of things or people and usually consist of a large body of data. They show how much one variable is affected by another. The relationship between the two variables is their **correlation**. The closer the data points come to making a straight line when plotted, the closer the correlation.

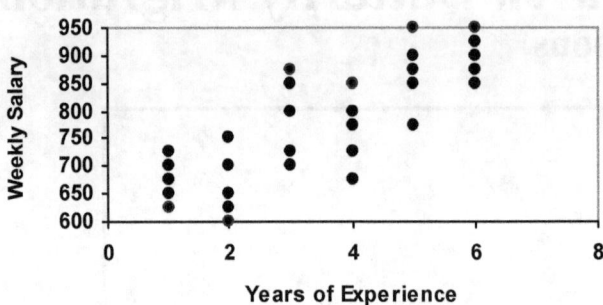

Stem and leaf plots are visually similar to line plots. The **stems** are the digits in the greatest place value of the data values, and the **leaves** are the digits in the next greatest place values. Stem and leaf plots are best suited for small sets of data and are especially useful for comparing two sets of data. The following is an example using test scores:

4	9
5	4 9
6	1 2 3 4 6 7 8 8
7	0 3 4 6 6 6 7 7 7 8 8 8 8
8	3 5 5 7 8
9	0 0 3 4 5
10	0 0

Histograms are used to summarize information from large sets of data that can be naturally grouped into intervals. The vertical axis indicates **frequency** (the number of times any particular data value occurs), and the horizontal axis indicates data values or ranges of data values. The number of data values in any interval is the **frequency of the interval**.

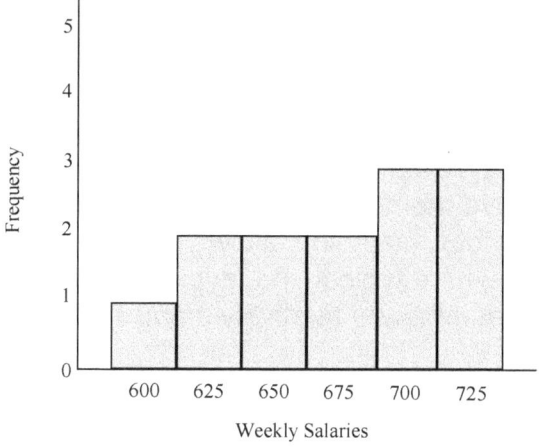

Bar graphs are similar to histograms. However, bar graphs are often used to convey information about categorical data where the horizontal scale represents a non-numeric attributes such as cities or years. Another difference is that the bars in bar graphs rarely touch. Bar graphs are also useful in comparing data about two or more similar groups of items.

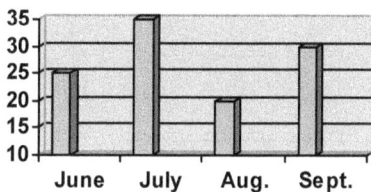

A **pie chart**, also known as a **circle graph**, is used to represent relative amounts of a whole.

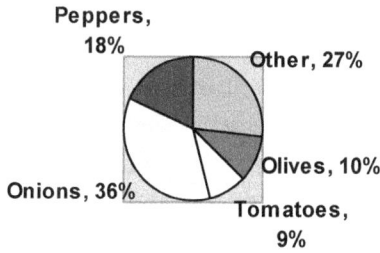

There are many graphical ways in which to represent data: line plots, line graphs, scatter plots, stem and leaf plots, histograms, bar graphs, pie charts, and pictographs.

A **line plot** organizes data in numerical order along a number line. An x is placed above the number line for each occurrence of the corresponding number. Line plots allow you to see at a glance a range of data and where typical and atypical data falls. They are generally used to summarize relatively small sets of data.

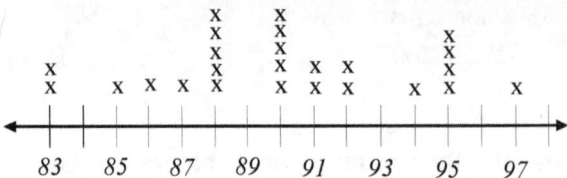

A **line graph** compares two variables, and each variable is plotted along an axis. A line graph highlights trends by drawing connecting lines between data points. They are particularly appropriate for representing data that varies continuously. Line graphs are sometimes referred to as **frequency polygons**.

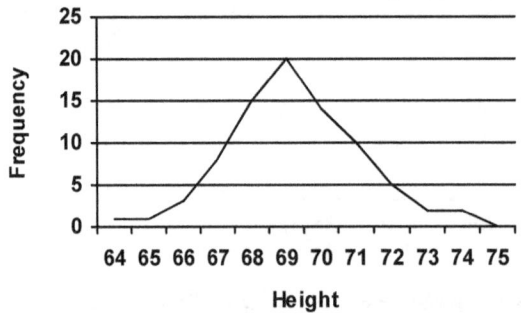

Graphs display data so that the data can be interpreted. Graphs are often used to see trends and predict future performance.

For example, the line graph on the next page graph the auto sales for a car dealership. The car dealership is able to see at a glance how many cars were sold in a particular month and which months tended to have the least and greatest sales. This information helps him to control his inventory, forecast his sales, and manage his staffing. He might also use the information to plan ways in which to boost sales in lagging months.

AUTO SALES

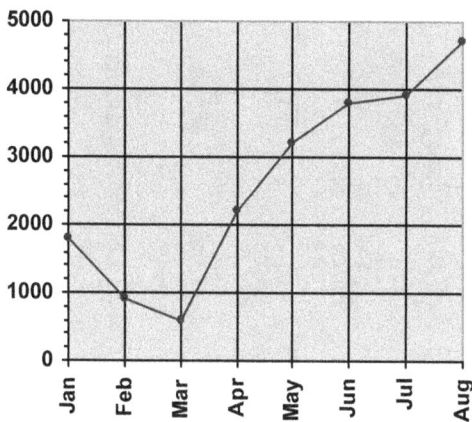

A **matrix** is an ordered set of numbers in rectangular form.

$$\begin{pmatrix} 0 & 3 & 1 \\ 4 & 2 & 3 \\ 1 & 0 & 2 \end{pmatrix}$$

Since this matrix has 3 rows and 3 columns, it is called a 3 x 3 matrix. The element in the second row, third column would be denoted as $3_{2,3}$.

A simple financial example of using a matrix to solve a problem follows:

A company has two stores. The income and expenses (in dollars) for the two stores, for three months, are shown in the matrices.

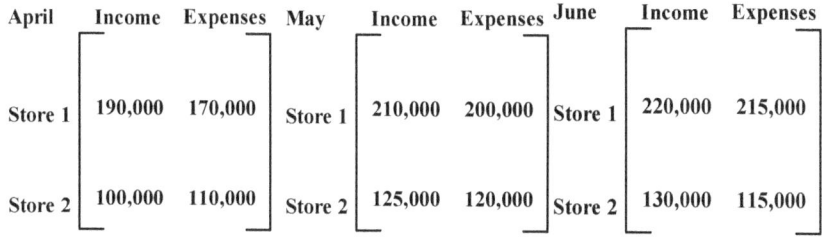

The owner wants to know what his first-quarter income and expenses were, so he adds the three matrices.

MATHEMATICS

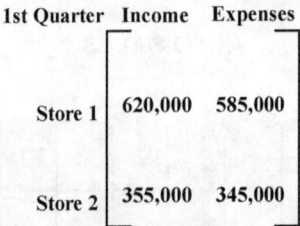

Then, to find the profit for each store:

Profit for Store 1 = $620,000 - $585,000 = $35,000
Profit for Store 2 = $355,000 - $345,000 = $10,000

An **algorithm** is a method of calculating; simply put, it can be multiplication, subtraction, or a combination of operations. When working with computers and calculators we employ **algorithmic thinking**, which means performing mathematical tasks by creating a sequential and often repetitive set of steps. A simple example would be to create an algorithm to generate the Fibonacci numbers utilizing the MR and M+ keys found on most calculators. The table below shows Entry made in the calculator, the value x seen in the display, and the value M contained in the memory.

Entry	ON/AC	1	M+	+	M+	MR	+	M+	MR	+	...
x	0	1	1	1	1	2	3	3	5	8	...
M	0	0	1	1	2	2	2	5	5	5	...

This eliminates the need to repeatedly enter required numbers.

Computers have to be programmed and many advanced calculators are programmable. A **program** is the steps of an algorithm that are entered into a computer or calculator.

The main advantage of using a program is that once the algorithm is entered, a result may be obtained by merely hitting a single keystroke to select the program, thereby eliminating the need to continually enter a large number of steps. Teachers find that programmable calculators are excellent for investigating "what if?" situations.

Using graphing calculators or computer software has many advantages. The technology is better able to handle large data sets, such as the results of a science experiment and it is much easier to edit and sort the data and change the style of the graph to find its best representation. Furthermore, graphing calculators also provide a tool to plot statistics.

0010 PROBLEMS

There are many example and practice problems in every section of this guide. A few more example problems are given below.

Exercise 1:

(a) Find the midpoint between (5, 2) and (-13, 4).

Using the Midpoint Formula:

$$\left(\frac{x_1+x_2}{2},\frac{y_1+y_2}{2}\right) = \left(\frac{5+(-13)}{2},\frac{2+4}{2}\right) = \left(\frac{-8}{2},\frac{6}{2}\right) = (-4,3)$$

(b) Find the value of x_1 so that (-3, 5) is the midpoint between (x_1, 6) and (-2, 4)

Using the Midpoint Formula:

$$(-3, 5) = \left(\frac{x_1+x_2}{2},\frac{y_1+y_2}{2}\right)$$

$$= \left(\frac{x_1+(-2)}{2},\frac{6+4}{2}\right)$$

$$= \left(\frac{x_1-2}{2},\frac{10}{2}\right)$$

$$= \left(\frac{x_1-2}{2},5\right)$$

Separate out the x value to determine x_1.

$$-3 = \frac{x_1-2}{2}$$
$$-6 = x_1 - 2$$
$$-4 = x_1$$

Exercise 2:

a) One line passes through the points (-4, -6) and (4, 6); another line passes through the points (-5, -4) and (3, 8). Are these lines parallel, perpendicular or neither?

Find the slopes.

$$m = \frac{y_2 - y_1}{x_2 - x_1}$$

$$m_1 = \frac{6-(-6)}{4-(-4)} = \frac{6+6}{4+4} = \frac{12}{8} = \frac{3}{2}$$

$$m_2 = \frac{8-(-4)}{3-(-5)} = \frac{8+4}{3+5} = \frac{12}{8} = \frac{3}{2}$$

Since the slopes are the same, the lines are parallel.

b) One line passes through the points (1, -3) and (0, -6); another line passes through the points (4, 1) and (-2, 3). Are these lines parallel, perpendicular or neither?

Find the slopes.

$$m = \frac{y_2 - y_1}{x_2 - x_1}$$

$$m_1 = \frac{-6-(-3)}{0-1} = \frac{-6+3}{-1} = \frac{-3}{-1} = 3$$

$$m_2 = \frac{3-1}{-2-4} = \frac{2}{-6} = -\frac{1}{3}$$

The slopes are negative reciprocals, so the lines are perpendicular.

c) One line passes through the points (-2, 4) and (2, 5); another line passes through the points (-1, 0) and (5, 4). Are these lines parallel, perpendicular or neither?

Find the slopes.

$$m = \frac{y_2 - y_1}{x_2 - x_1}$$

$$m_1 = \frac{5-4}{2-(-2)} = \frac{1}{2+2} = \frac{1}{4}$$

$$m_2 = \frac{4-0}{5-(-1)} = \frac{4}{5+1} = \frac{4}{6} = \frac{2}{3}$$

Since the slopes are not the same, the lines are not parallel. Since they are not negative reciprocals, they are not perpendicular, either. Therefore, the answer is "neither."

0011 MODELS

Exercise:

For 2000 through 2005, the consumption of a certain product sweetened with sugar, as a percent, $f(t)$, of the total consumption of the product, can be modeled by:

$$f(t) = 75 + 37.25(0.615)^t$$

where $t = 2$ represents 2000.

(a) Find a model for the consumption of the product sweetened with non-sugar sweeteners as a percent, $g(t)$, of the total consumption of the product.

Since 100% represents the total consumption of the product, the model can be found by subtracting the model for sugar-sweetened product from 100:

$$\begin{aligned} g(t) &= 100 - (75 + 37.25(0.615)^t \\ &= 100 - 75 - 37.25(0.615)^t \\ &= 25 - 37.25(0.615)^t \end{aligned}$$

(b) Sketch the graphs of f and g. Does the consumption of one type of product seem to be stabilizing compared to the other product? Explain.

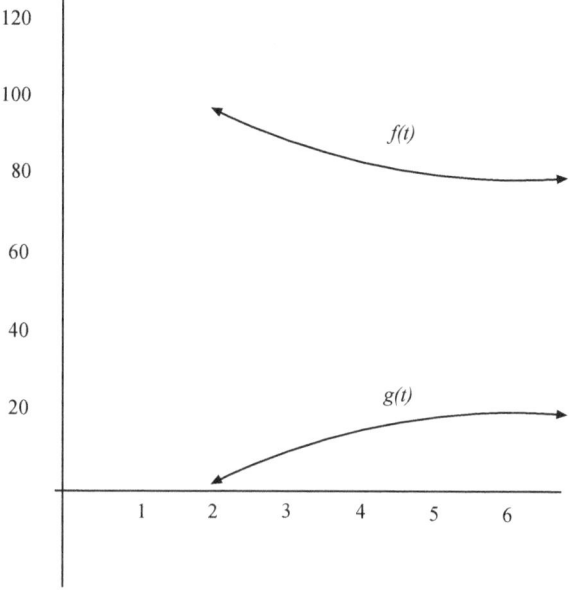

Yes, the consumption of the product sweetened with sugar (represented by $f(t)$) is decreasing less and less each year.

(c) Sketch the graph of $f(x) = 2^x$. Does it have an x-intercept? What does this tell you about the number of solutions of the equation $2^x = 0$? Explain.

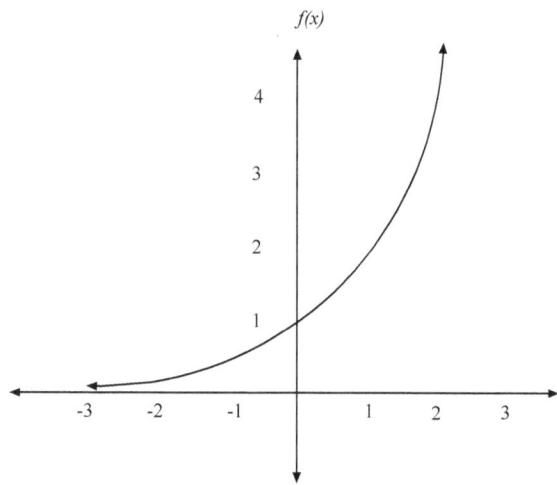

No, there is no solution. The solutions of $2^x = 0$ are the x-intercepts of $y = 2^x$.

0012   PROOFS

For examples of geometric proofs see pages 42-46. For examples of trigonometric proofs see pages 76-78.

TEACHER CERTIFICATION STUDY GUIDE

CURRICULUM AND INSTRUCTION

TEACHING METHODS - The art and science specific for high school mathematics

Some commonly used teaching techniques and tools are described below along with links to further information. The links provided provide a wealth of instructional ideas and materials. You should consider joining The National Council of Teachers of Mathematics as they have many ideas in their journals about pedagogy and curriculum standards and publish professional books that are useful. You can write to them at 1906 Association Drive, Reston, VA 20191-1593. You can also order a starter kit from them for $9 that includes 3 recent journals by calling 800-235-7566 or writing e-mailorders@nctm.org.

> *A couple of resources for students to use at home:*
> *http://www.algebra.com/, http://www.mathsisfun.com/algebra/index.html*
> *and http://www.purplemath.com/. A helpful .pdf guide for parents is*
> *available at:*
> *http://my.nctm.org/ebusiness/ProductCatalog/product.aspx?ID=12931*
> A good website for understanding the causes of and how to prevent "math anxiety:"
> http://www.mathgoodies.com/articles/math_anxiety.html

1. Classroom warm-up: Engage your students as soon as they walk in the door: provide an interesting short activity each day. You can make use of thought-provoking questions and puzzles. Also use relevant puzzles specific to topics you may be covering in your class. The following websites provide some ideas:
   http://www.math-drills.com/?gclid=CP-P0dzenJICFRSTGgodNjG0Zw
   http://www.mathgoodies.com/games/
   http://mathforum.org/k12/k12puzzles/
   http://mathforum.org/pow/other.html

2. Real life examples: Connect math to other aspects of your students' lives by using examples and data from the real world whenever possible. It will not only keep them engaged, it will also help answer the perennial question "Why do we have to learn math?" Online resources to get you started:

   http://chance.dartmouth.edu/chancewiki/index.php/Main_Page has some interesting real-world probability problems (such as, "Can statistics determine if Robert Clemens used steroids?")
   http://www.mathnotes.com/nos_index.html has all kinds of links between math and the real world suitable for high school students
   http://www.nssl.noaa.gov/edu/ideas/ uses weather to teach math

MATHEMATICS

http://standards.nctm.org/document/eexamples/index.htm#9-12 Using Graphs, Equations, and Tables to Investigate the Elimination of Medicine from the Body: Modeling the Situation
http://mathforum.org/t2t/faq/election.html Election math in the classroom
http://www.education-world.com/a_curr/curr148.shtml offers examples of real-life problems such as calculating car payments, saving and investing, the world of credit cards, and other finance problems.
http://score.kings.k12.ca.us/real.world.html is a website connecting math to real jobs, elections, NASA projects, etc.

3. Graphing and spreadsheets for enhancing math learning:
   http://www.cvgs.k12.va.us/digstats/
   http://score.kings.k12.ca.us/standards/probability.html for graphing and statistics.
   http://www.microsoft.com/education/solving.mspx for using spreadsheets and to solve polynomial problems.

4. Use technology - manipulatives, software and interactive online activities that can help all students learn, particularly those oriented more towards visual and kinesthetic learning. Here are some websites:
   http://illuminations.nctm.org/ActivitySearch.aspx has games for grades 9-12 that can be played against the computer or another student.
   http://nlvm.usu.edu/ The National Library of Virtual Manipulatives has resources for all grades on numbers and operations, algebra, geometry, probability and measurement.
   http://mathforum.org/pow/other.html has links to various math challenges, manipulatives and puzzles.
   http://www.etacuisenaire.com/algeblocks/algeblocks.jsp Algeblocks are blocks that utilize the relationship between algebra and geometry.

5. Word problem strategies- the hardest thing to do is take the English and turn it into math but there are 6 key steps to teach students how to solve word problems:

   a. The problem will have **key words** to suggest the type of operation or operations to be performed to solve the problem. For example, words such as "altogether" or "total" imply addition while words such as "difference" or "How many more?" imply subtraction.
   b. **Pictures or Concrete Materials**: Math is very abstract; it is easier to solve a problem using pictures or concrete materials to illustrate the problem. Pictures and concrete materials allow the students to manipulate the material to solve the problem with trial and error. Model drawing pictures and using concrete materials to solve word problems.

c. ***Use Logic***: Ask your students if their answers make sense. Get them used to using the process of deduction. Model the deduction process for them to decide on the answer to a word problem. For example, in solving a problem such as: Two consecutive numbers have a sum of 91. What are the numbers? If the student arrives at an answer of 44 and 45 it is obvious that there was an error in the equation used or calculation since 44 and 45 are consecutive but don't add up to 91. Let x = the 1st number and (x+1) = the 2nd number, so that x + (x+1) =91 and 2x +1 =91, then 2x=90, x=45 and x+1=46. The answer is 45 and 46.
d. ***Eliminate the possibilities and look for patterns or work the problem backwards***
e. ***Guess the Answer***: Students should guess an approximate answer that makes sense based on the problem. For example, if the student knows the word problem implies addition, they should recognize that the answer must be greater than the numbers in the problem. Often students are afraid of guessing because they don't want to get the wrong answer but encourage your students to guess and then double check the answer to see if it works. If the answer is incorrect, the student can try another strategy for finding the answer.
f. ***Make a Table***: Selecting relevant information from a word problem and organizing the data is very helpful in solving word problems. Often students become confused because there are too many numbers and/or variables.

http://www.purplemath.com/modules/translat.htm,
http://math.about.com/library/weekly/aa071002a.htm and
http://www.onlinemathlearning.com/algebra-word-problems.html are great resources for students to solve word problems.

6. Mental math practice
Give students regular practice in doing mental math. The following websites offer many mental calculation tips and strategies:
http://www.cramweb.com/math/index.htm
http://mathforum.org/k12/mathtips/mathtips.html

Because frequent calculator use tends to deprive students of a sense of numbers and an ability to calculate on their own, they will often approach a sequence of multiplications and divisions the hard way. For instance, asked to calculate 770 x 36/ 55, they will first multiply 770 and 36 and then do a long division with the 55. They fail to recognize that both 770 and 55 can be divided by 11 and then by 5 to considerably simplify the problem. Give students plenty of practice in multiplying and dividing a sequence of integers and fractions so they are comfortable with canceling top and bottom terms.

7. Math language
   Math vocabulary help is available for high school students on the web:

   http://www.amathsdictionaryforkids.com/ is a colorful website math dictionary
   http://www.math.com/tables/index.html is a math dictionary in English and Spanish

ERROR ANALYSIS

A simple method for analyzing student errors is to ask how the answer was obtained. The teacher can then determine if a common error pattern has resulted in the wrong answer. There is a value to having the students explain how the arrived at the correct as well as the incorrect answers.

Many errors are due to simple **carelessness**. Students need to be encouraged to work slowly and carefully. They should check their calculations by redoing the problem on another paper, not merely looking at the work. Addition and subtraction problems need to be written neatly so the numbers line up. Students need to be careful regrouping in subtraction. Students must write clearly and legibly, including erasing fully. Use estimation to ensure that answers make sense.

Many students' computational skills exceed their **reading** level. Although they can understand basic operations, they fail to grasp the concept or completely understand the question. Students must read directions slowly.
Fractions are often a source of many errors. Students need to be reminded to use common denominators when adding and subtracting and to always express answers in simplest terms. Again, it is helpful to check by estimating.

The most common error that is made when working with **decimals** is failure to line up the decimal points when adding or subtracting or not moving the decimal point when multiplying or dividing. Students also need to be reminded to add zeroes when necessary. Reading aloud may also be beneficial. Estimation, as always, is especially important.

Students need to know that it is okay to make mistakes. The teacher must keep a positive attitude, so they do not feel defeated or frustrated.

TEACHER CERTIFICATION STUDY GUIDE

THE ART OF TEACHING - PEDAGOGICAL PRINCIPLES
Maintain a supportive, non-threatening environment

The key to success in teaching goes beyond your mathematical knowledge and the desire to teach. Though important, knowledge and desire alone do not make you a good teacher. Being able to connect with your students is vital: learn their names immediately, have a seating chart the first day (even if you intend to change it) and learn about your students -their hobbies, phone number, parent's names, what they like and dislike about school and learning and math. Keep this information on each student and learn it: adapt your lessons, how challenging they are and what other resources you may need to accommodate your students' individual strengths and weaknesses.

Learn to see math as your students see it. If you aren't able to connect with your students, no matter how well your lessons are and how well you know the material, you won't inspire them to learn math from you. As you expect respect, you must give respect and as you expect their attention, they also need your attention and understanding. Talk to them with the same tone of voice as you would an adult, not in a tone that makes them feel like children. Look your students in the eye when you talk to them and encourage questions and comments. Take advantage of teachable moments and explain the rationale behind math rules.

Demonstrate respect, care and trust toward every student; assume the best. This does not mean becoming "friends" with your students or you will have problems with discipline. You can be kind and firm at the same time. Have a fair and clear grading and discipline system that is posted, reviewed and made clear to your students. Consistency, structure and fairness are essential to earning their trust in you as a teacher. Always admit your mistakes and be available to your students certain days after school. Finally, demonstrate your care for them and your love of math and you will be a positive influence on their learning.

Below are websites to help make your teaching more effective and fun:

1. **Teachers Helping Teachers** has several resources for high school mathematics.
2. **Math Resources for Teachers** – resources for grades 7 - 10
3. **Math is Marvelous Web Site** - is a fascinating website on the history of geometry
4. **Math Archives K-12** resources for lesson plans and software
5. **http://www.edhelper.com/algebra.htm** covers Algebra I & II
6. **Math Goodies** interactive lessons, worksheets and homework help
7. **Multicultural Lessons** an interesting site with lessons on Babylonian Square Roots, Chinese Fraction Reducing, Egyptian multiplication, etc.
8. **http://www.goenc.com/** resources and professional development for teachers
9. **Math and Reading Help** a guide to math, reading, homework help, tutoring and earning a high school diploma

MATHEMATICS

10. **Purple Math.com** all about Algebra, lessons, help for students and lots of other resources
11. http://library.thinkquest.org/20991/home.html this site has Algebra, Geometry and Pre-calc/Calculus
12. http://www.math.com/ this site has Algebra, Geometry, Trigonometry, and Calculus, plus homework help
13. http://www.math.armstrong.edu/MathTutorial/index.html a tutorial in algebra
14. http://www.wtamu.edu/academic/anns/mps/math/mathlab/beg_algebra/index.htm this site is helpful for those beginning Algebra or for a refresher
15. **Math Complete** this radicals, quadratics, linear equation solvers
16. **Math Tutor - PEMDAS & Integers** fractions, integers, information for parents and teachers
17. **Matrix Algebra** all about matrix operations and applications
18. **Mr. Stroh's Algebra Site** help for Algebra I & II
19. **Polynomials and Polynomial Functions** everything from factoring, to graphing, finding rational zeros and multiplying, adding and subtracting polynomials
20. **Quadratic Formula** all about quadratics
21. **Animated Pythagorean Theorem** a fun an animated proof of the Pythagorean Theorem
22. **Brunnermath - Interactive Activities** general math, Algebra, Geometry, Trigonometry, Statistics, Calculus, using Calculators
23. **CoolMath4Kids Geometry** creating art with math and geometry lessons
24. **The Curlicue Fractal** The curlicue fractal is an exceedingly easy-to-make but richly complex pattern using trigonometry and calculus to create fascinating shapes
25. **Euclid's Elements Interactive** Euclid's *Elements* form one of the most beautiful and influential works of science in the history of humankind.
26. **Howe-Two Free Software** software solutions for mathematics instruction
27. http://regentsprep.org/regents/math/math-topic.cfm?TopicCode=syslin Systems of equations lessons and practice
28. http://www.sparknotes.com/math/algebra1/systemsofequations/problems3.rhtml Word problems system of equations
29. http://math.about.com/od/complexnumbers/Complex_Numbers.htm Several complex number exercise pages
30. http://regentsprep.org/Regents/math/ALGEBRA/AE3/PracWord.htm practice with Systems of inequalities word problems
31. http://regentsprep.org/regents/Math/solvin/PSolvIn.htm solving inequalities
32. http://www.wtamu.edu/academic/anns/mps/math/mathlab/beg_algebra/beg_alg_tut18_ineq.htm Inequality tutorial, examples, problems
33. http://www.wtamu.edu/academic/anns/mps/math/mathlab/beg_algebra/beg_alg_tut24_ineq.htm Graphing linear inequalities tutorial

34. **http://www.wtamu.edu/academic/anns/mps/math/mathlab/col_algebra/col_alg_tut17_quad.htm** Quadratic equations tutorial, examples, problems
35. **http://regentsprep.org/Regents/math/math-topic.cfm?TopicCode=factor** Practice factoring
36. **http://www.wtamu.edu/academic/anns/mps/math/mathlab/col_algebra/col_alg_tut37_syndiv.htm** Synthetic division tutorial
37. **http://www.tpub.com/math1/10h.htm** Synthetic division Examples and problems

DEVELOPMENTAL PSYCHOLOGY AND TEACHING MATHEMATICS- things you may not know about your students:

Studies show that health matters more than gender or social status when it comes to learning. Healthy girls and boys do equally well on most cognitive tasks. Boys perform better at manipulating shapes and analyzing and girls perform better on processing speed and motor dexterity. No differences have been measured in calculation ability, meaning girls and boys have an equal aptitude for mathematics.

The following was written by Jay Giedd, M.D. is a practicing Child and Adolescent Psychiatrist and Chief of Brain Imaging at the Child Psychiatry Branch of the National Institute of Mental Health:

http://nihrecord.od.nih.gov/newsletters/2005/08_12_2005/story04.htm

"The most surprising thing has been how much the teen brain is changing. By age six, the brain is already 95 percent of its adult size. But the gray matter, or thinking part of the brain, continues to thicken throughout childhood as the brain cells get extra connections, much like a tree growing extra branches, twigs and roots...In the frontal part of the brain, the part of the brain involved in judgment, organization, planning, strategizing -- those very skills that teens get better and better at -- this process of thickening of the gray matter peaks at about age 11 in girls and age 12 in boys, roughly about the same time as puberty. After that peak, the gray matter thins as the excess connections are eliminated or pruned...

Contrary to what most parents have thought at least once, "teens really do have brains," quipped Dr. Jay Giedd, NIMH intramural scientist, in a lecture on the "Teen Brain Under Construction." His talk was the kick-off event for the recent NIH Parenting Festival. Giedd said scientists have only recently learned more about the trajectories of brain growth. One of the findings he discussed showed the frontal cortex area — which governs judgment, decision-making and impulse control — doesn't fully mature until around age 25. "That really threw us," he said. "We used to joke about having to be 25 to rent a car, but there's tons of data from insurance reports [showing] that 24-year-olds are costing them more than 44-year-olds."

So why is that? "It must be behavior and impulse control," he said. "Whatever these changes are, the top 10 bad things that happen to teens involve emotion and behavior." Physically, Giedd said, the teen years and early 20s represent an incredibly healthy time of life, in terms of cancer, heart disease and other serious illnesses. But with accidents as the leading cause of death in adolescents, and suicide following close behind, "this isn't a great time emotionally and psychologically. This is the great paradox of adolescence: right at the time you should be on the top of your game, you're not."

The next step in Giedd's research, he said, is to learn more about what influences brain growth, for good or bad. "Ultimately, we want to use these findings to treat illness and enhance development."

One of the things scientists have come to understand, though, is that parents do have something to do with their children's brain development.

"From imaging studies, one of the things that seems intriguing is this notion of modeling...that the brain is pretty adept at learning by example," he said. "As parents, we teach a lot when we don't even know we're teaching, just by showing how we treat our spouses, how we treat other people, what we talk about in the car on the way home...things that a parent says in the car can stick with them for years. They're listening even though it may appear they're not."

What can we do to change our kids? "Well, start with yourself in terms of what you show by example," Giedd concluded.

Maybe the parts of the brain performing geometry are different from the parts doing algebra. There is no definitive research to answer that question yet, but it is obviously what researchers are looking for.

Time-Lapse Imaging Tracks Brain Maturation from ages 5 to 20
Constructed from MRI scans of healthy children and teens, the time-lapse "movie", from which the above images were extracted, compresses 15 years of brain development (ages 5–20) into just a few seconds.

To view in color, go to the website below: Red indicates **more** gray matter, **blue less** gray matter. Gray matter wanes in a back-to-front wave as the brain matures and neural connections are pruned.
Source: Paul Thompson, Ph.D. UCLA Laboratory of Neuroimaging
http://www.loni.ucla.edu/%7Ethompson/DEVEL/PR.html

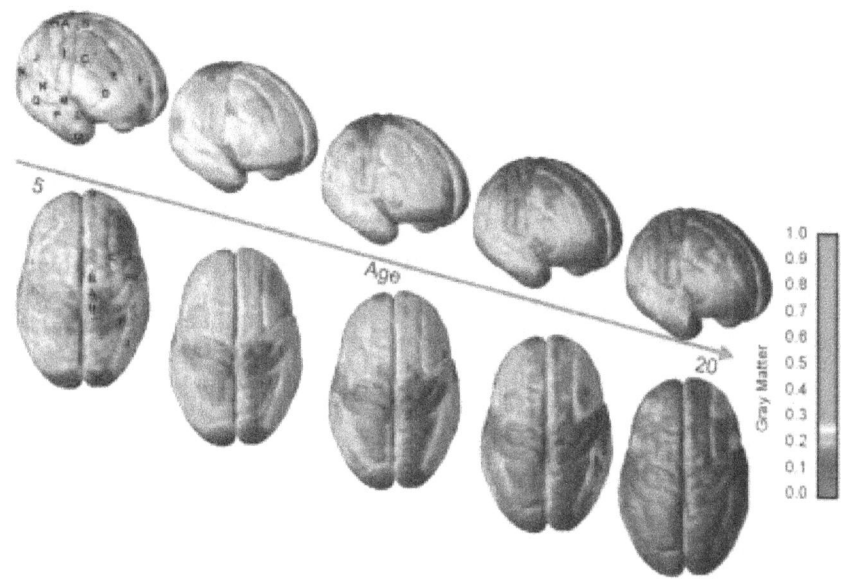

What are the implications of this fascinating study for teachers? It's unreasonable to expect teens to have adult levels of organizational skills or decision-making before their brains have completely developed. In teens, the frontal lobe, or the executive of the brain is what handles organizing, decision making, emotions, attending, shifting attention, planning and making strategies and it is not fully developed until the early to mid-twenties.

Perhaps since certain parts of the brain develop sooner than others, subjects should be taught in a different order. Until we know more, just understanding that the parts of teen brains related to decision making and emotions are still developing through the early 20's is important, and **that stressful situations lead to diminished ability to made good judgments.** *For some children, just being called on in class is stressful. At this age, social relationships become very important and* **teachers need to be sensitive to this aspect of teen development as it relates to stress and decision-making.** *The immaturity of this part of the teen brain might explain why the teen crash rate is 4 times that of adults...*

TEACHER CERTIFICATION STUDY GUIDE

ANSWER KEY TO PRACTICE PROBLEMS

0001 ALGEBRA AND NUMBER THEORY

Page 10

Question #1 $\dfrac{8x+36}{(x-3)(x+7)}$

Question #2 $\dfrac{25a^2 + 12b^2}{20a^4 b^5}$

Question #3 $\dfrac{2x^2 + 5x - 21}{(x-5)(x+5)(x+3)}$

Page 13

Question #1 It takes Curly 15 minutes to paint the elephant alone
Question #2 The original number is 5/15
Question #3 The car was traveling at 68mph and the truck was traveling at 62mph

Page 14

Question #1 $C = \dfrac{5}{9}F - \dfrac{160}{9}$

Question #2 $b = \dfrac{2A - 2h^2}{h}$

Question #3 $n = \dfrac{360 + S}{180}$

0004 TRIGONOMETRY

Page 77

Question #1
$$\cot\theta = \dfrac{x}{y}$$
$$\dfrac{x}{y} = \dfrac{x}{r} \times \dfrac{r}{y} = \dfrac{x}{y} = \cot\theta$$

Question #2
$$1 + \cot^2\theta = \csc^2\theta$$
$$\dfrac{y^2}{y^2} + \dfrac{x^2}{y^2} = \dfrac{r^2}{y^2} = \csc^2\theta$$

MATHEMATICS

0005 FUNCTIONS

Page 101

Question #2 $x = 9$
Question #3 $x = 3, x = 6$
Question #4 1

Page 103

Question #1

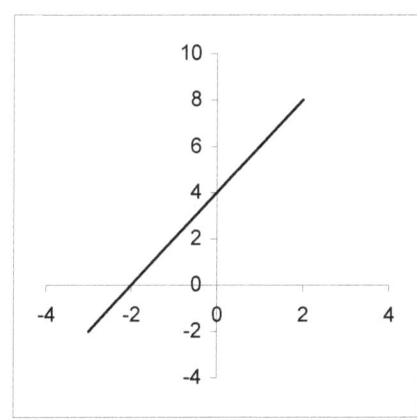

Question #2

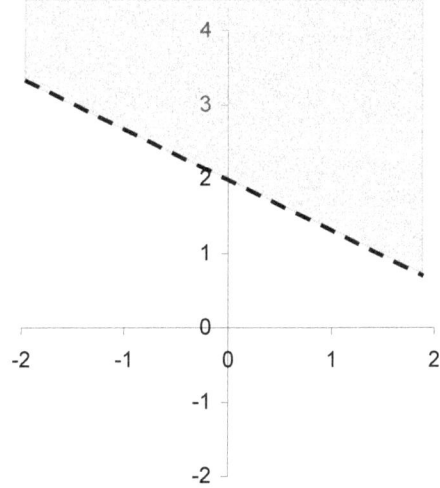

Question #3

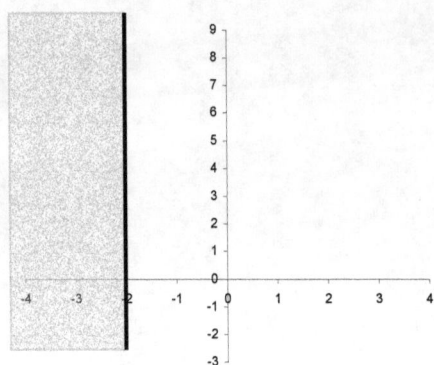

0006   CALCULUS

Page 105

    Question #1 $5\pi^2$
    Question #2   1

Page 106

    Question #1 ∞
    Question #2   -1

Page 122

    Question #1 $g(x) = -20\sin x + c$
    Question #2 $g(x) = \pi \sec x + c$

Page 123

    Question #1 $t(0) = -24$ m/sec
    Question #2 $t(4) = 24$ m/sec

0004   MATRIX ALGEBRA

Page 153

    Question #1 $\begin{pmatrix} 11 & 7 \\ 11 & 1 \end{pmatrix}$

    Question #2 $\begin{pmatrix} 0 & 3 \\ -10 & 13 \\ 5 & 0 \end{pmatrix}$

Question #3 $\begin{pmatrix} -4 & 0 & -2 \\ 2 & 4 & -8 \end{pmatrix}$

Question #4 $\begin{pmatrix} 18 \\ 34 \\ 32 \end{pmatrix}$

Question #5 $\begin{pmatrix} -12 & 16 \\ -4 & -2 \\ 0 & 6 \end{pmatrix}$

Page 155

Question #1 $\begin{pmatrix} -7 & 0 \\ -12 & -3 \\ 6 & 5 \end{pmatrix}$

Question #2 $\begin{pmatrix} 3 & \frac{47}{4} \\ 3 & 7 \end{pmatrix}$

Question #3   a = -5   b = -1   c = 0   d = -4   e = 0   f = -10

Page 156

Question #1 $\begin{pmatrix} -15 & 25 \\ -1 & -13 \end{pmatrix}$

Question #2 $\begin{pmatrix} 5 & -5 & -10 \\ 5 & 5 & 0 \\ 1 & 8 & 7 \\ -9 & 13 & 22 \end{pmatrix}$

Page 157

Question #1 $\begin{pmatrix} x \\ y \end{pmatrix} = \begin{pmatrix} 3 \\ 1 \end{pmatrix}$

Question #2 $\begin{pmatrix} x \\ y \\ z \end{pmatrix} = \begin{pmatrix} 4 \\ 4 \\ 1 \end{pmatrix}$

0009   DISCRETE MATHEMATICS

Page 161

Question #1 $S_5 = 75$

Question #2 $S_n = 28$

Question #3 $S_n = -\dfrac{-31122}{15625} \approx^- 1.99$

TEACHER CERTIFICATION STUDY GUIDE

Sample Test

Directions: Read each item and select the best response.

1. Find the LCM of 27, 90 and 84.
   (Easy)(Skill 0001)

   A) 90
   B) 3780
   C) 204120
   D) 1260

2. Solve for x by factoring
   $2x^2 - 3x - 2 = 0$.
   (Average Rigor) (Skill 0001)

   A) x = (-1,2)
   B) x = (0.5,-2)
   C) x=(-0.5,2)
   D) x=(1,-2)

3. What would be the total cost of a suit for $295.99 and a pair of shoes for $69.95 including 6.5% sales tax?
   (Average Rigor) (Skill 0001)

   A) $389.73
   B) $398.37
   C) $237.86
   D) $315.23

4. Which of the following is always composite if x is odd, y is even, and both x and y are greater than or equal to 2?
   (Average Rigor)(Skill 0001)

   A) $x+y$
   B) $3x+2y$
   C) $5xy$
   D) $5x+3y$

5. What is the smallest number that is divisible by 3 and 5 and leaves a remainder of 3 when divided by 7?
   (Average Rigor)(Skill 0001)

   A) 15
   B) 18
   C) 25
   D) 45

6. Given the series of examples below, what is 5¢4?
   (Average Rigor)(Skill 0001)

   4¢3=13   7¢2=47
   3¢1=8   1¢5=-4

   A) 20
   B) 29
   C) 1
   D) 21

MATHEMATICS

7. Which of the following is a factor of the expression $9x^2 + 6x - 35$?
   (Rigorous)(Skill 0001)

   A) 3x-5
   B) 3x-7
   C) x+3
   D) x-2

8. Which of the following is a factor of $6 + 48m^3$
   (Rigorous) (Skill 0001)

   A) (1 + 2m)
   B) (1 - 8m)
   C) (1 + m - 2m)
   D) (1 - m + 2m)

9. Which of the following is incorrect?
   (Rigorous) (Skill 0001)

   A) $(x^2 y^3)^2 = x^4 y^6$
   B) $m^2(2n)^3 = 8m^2 n^3$
   C) $(m^3 n^4)/(m^2 n^2) = mn^2$
   D) $(x + y^2)^2 = x^2 + y^4$

10. Solve for x: $18 = 4 + |2x|$
    (Rigorous) (Skill 0001)

    A) $\{-11, 7\}$
    B) $\{-7, 0, 7\}$
    C) $\{-7, 7\}$
    D) $\{-11, 11\}$

11. Find the surface area of a box which is 3 feet wide, 5 feet tall, and 4 feet deep.
    (Easy)(Skill 0002)

    A) 47 sq. ft.
    B) 60 sq. ft.
    C) 94 sq. ft
    D) 188 sq. ft.

12. The term "cubic feet" indicates which kind of measurement?
    (Average Rigor)(Skill 0002)

    A) Volume
    B) Mass
    C) Length
    D) Distance

13. Given a 30 meter x 60 meter garden with a circular fountain with a 5 meter radius, calculate the area of the portion of the garden not occupied by the fountain.
    (Average Rigor)(Skill 0002)

    A) 1721 m²
    B) 1879 m²
    C) 2585 m²
    D) 1015 m²

MATHEMATICS

14. Find the height of a box with surface area of 94 sq. ft. with a width of 3 feet and a depth of 4 feet.
    (Average Rigor)(Skill 0002)

    A) 3 ft.
    B) 4 ft.
    C) 5 ft
    D) 6 ft.

15. If the area of the base of a cone is tripled, the volume will be
    (Rigorous)(Skill 0002)

    A) the same as the original
    B) 9 times the original
    C) 3 times the original
    D) 3π times the original

16. Find the area of the figure pictured below.
    (Rigorous)(Skill 0002)

    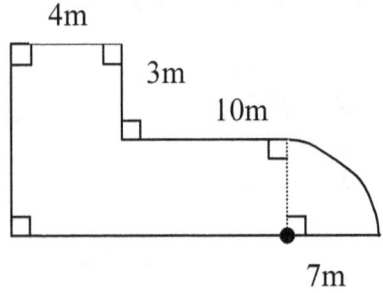

    A) 136.47 m²
    B) 148.48 m²
    C) 293.86 m²
    D) 178.47 m²

17. Compute the area of the shaded region, given a radius of 5 meters. 0 is the center.
    (Rigorous)(Skill 0002)

    A) 7.13 cm²
    B) 7.13 m²
    C) 78.5 m²
    D) 19.63 m²

    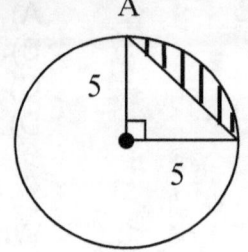

18. The length of a picture frame is 2 inches greater than its width. If the area of the frame is 143 square inches, what is its width?
    (Rigorous)(Skill 0002)

    A) 11 inches
    B) 13 inches
    C) 12 inches
    D) 10 inches

19. Determine the area of the shaded region of the trapezoid in terms of x and y.
    (Rigorous)(Skill 0002)

    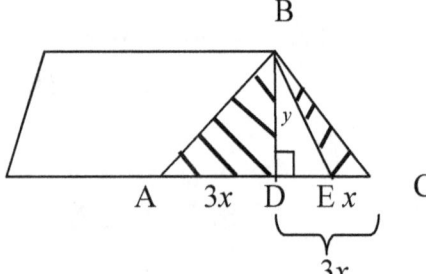

    A) $4xy$
    B) $2xy$
    C) $3x^2y$
    D) There is not enough information given.

20. When you begin by assuming the conclusion of a theorem is false, then show that through a sequence of logically correct steps you contradict an accepted fact, this is known as
(Easy)(Skill 0003)

A) inductive reasoning
B) direct proof
C) indirect proof
D) exhaustive proof

21. Which theorem can be used to prove $\triangle BAK \cong \triangle MKA$?
(Average Rigor)(Skill 0003)

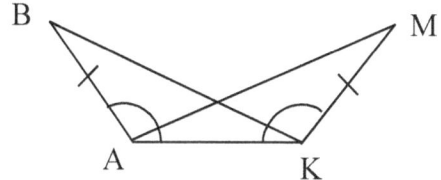

A) SSS
B) ASA
C) SAS
D) AAS

22. Choose the correct statement concerning the median and altitude in a triangle.
(Average Rigor)(Skill 0003)

A) The median and altitude of a triangle may be the same segment.
B) The median and altitude of a triangle are always different segments.
C) The median and altitude of a right triangle are always the same segment.
D) The median and altitude of an isosceles triangle are always the same segment.

23. Compute the distance from (-2, 7) to the line $x = 5$.
(Average Rigor)(Skill 0003)

A) -9
B) -7
C) 5
D) 7

24. Which of the following statements about a trapezoid is incorrect? (Average Rigor)(Skill 0003)

    A) It has one pair of parallel sides
    B) The parallel sides are called bases
    C) If the two bases are the same length, the trapezoid is called isosceles
    D) The median is parallel to the bases

25. Given $K(-4, y)$ and $M(2, -3)$ with midpoint $L(x, 1)$, determine the values of x and y. (Rigorous)(Skill 0003)

    A) $x = -1, y = 5$
    B) $x = 3, y = 2$
    C) $x = 5, y = -1$
    D) $x = -1, y = -1$

26. Which equation represents a circle with a diameter whose endpoints are $(0, 7)$ and $(0, 3)$? (Rigorous)(Skill 0003)

    A) $x^2 + y^2 + 21 = 0$
    B) $x^2 + y^2 - 10y + 21 = 0$
    C) $x^2 + y^2 - 10y + 9 = 0$
    D) $x^2 - y^2 - 10y + 9 = 0$

27. What is the degree measure of an interior angle of a regular 10 sided polygon? (Rigorous)(Skill 0003)

    A) 18°
    B) 36°
    C) 144°
    D) 54°

28. What is the measure of minor arc AD, given measure of arc PS is 40° and $m < K = 10°$? (Rigorous)(Skill 0003)

    A) 50°
    B) 20°
    C) 30°
    D) 25°

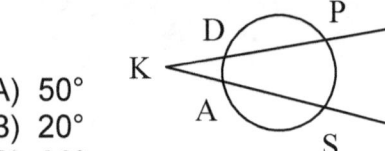

29. Find the length of the major axis of $x^2 + 9y^2 = 36$. (Rigorous)(Skill 0003)

    A) 4
    B) 6
    C) 12
    D) 8

30. The cosine function is defined as (Easy)(Skill 0004)

    A) base/hypotenuse
    B) perpendicular/hypotenuse
    C) hypotenuse/base
    D) perpendicular/base

31. The following is a Pythagorean identity: (Easy)(Skill 0004)

   A) $\sin^2\theta - \cos^2\theta = 1$
   B) $\sin^2\theta + \cos^2\theta = 1$
   C) $\cos^2\theta - \sin^2\theta = 1$
   D) $\cos^2\theta + \tan^2\theta = 1$

32. Determine the measures of angles A and B. (Average Rigor)(Skill 0004)

   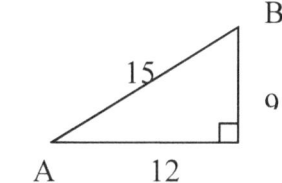

   A) A = 30°, B = 60°
   B) A = 60°, B = 30°
   C) A = 53°, B = 37°
   D) A = 37°, B = 53°

33. Which expression is not identical to sinx? (Average Rigor)(Skill 0004)

   A) $\sqrt{1-\cos^2 x}$
   B) $\tan x \cos x$
   C) $1/\csc x$
   D) $1/\sec x$

34. Determine the rectangular coordinates of the point with polar coordinates (5, 60°). (Average Rigor)(Skill 0004)

   A) (0.5, 0.87)
   B) (-0.5, 0.87)
   C) (2.5, 4.33)
   D) (25, 150°)

35. Which expression is equivalent to $1-\sin^2 x$? (Rigorous)(Skill 0004)

   A) $1-\cos^2 x$
   B) $1+\cos^2 x$
   C) $1/\sec x$
   D) $1/\sec^2 x$

36. For an acute angle x, sinx = 3/5. What is cotx? (Rigorous)(Skill 0004)

   A) 5/3
   B) 3/4
   C) 1.33
   D) 1

37. What is the equation of the graph shown below? (Easy)(Skill 0005)

   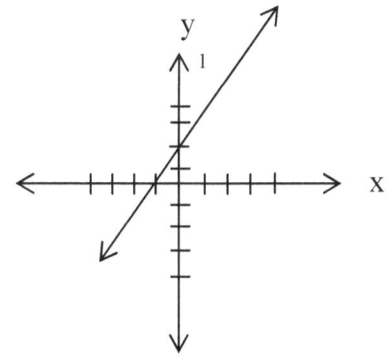

   A) $2x + y = 2$
   B) $2x - y = -2$
   C) $2x - y = 2$
   D) $2x + y = -2$

38. Solve for v_0: $d = at(v_t - v_0)$
(Average Rigor)(Skill 0005)

A) $v_0 = atd - v_t$
B) $v_0 = d - atv_t$
C) $v_0 = atv_t - d$
D) $v_0 = (atv_t - d)/at$

39. Given $f(x) = 3x - 2$ and $g(x) = x^2$, determine $g(f(x))$.
(Average Rigor)(Skill 0005)

A) $3x^2 - 2$
B) $9x^2 + 4$
C) $9x^2 - 12x + 4$
D) $3x^3 - 2$

40. Which graph represents the solution set for $x^2 - 5x > -6$?
(Average Rigor)(Skill 0005)

A) ←—○——○—→
     -2 0 2

B) ←○——————○→
     -3 0 3

C) ←—○———○—→
     -2 0 2

D) ←————○○—→
     -3 0 2 3

41. Solve for x:
$6(x + 2) - 5 = 21 - x$.
(Average Rigor)(Skill 0005)

A) 3
B) 2
C) 5
D) 0

42. How does the function $y = x^3 + x^2 + 4$ behave from $x = 1$ to $x = 3$?
(Average Rigor) (Skill 0005)

A) increasing, then decreasing
B) increasing
C) decreasing
D) neither increasing nor decreasing

43. Which graph represents the equation of $y = x^2 + 3x$?
(Average Rigor) (Skill 0005)

A)   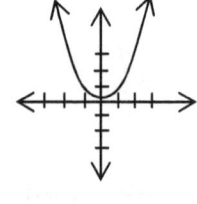   B)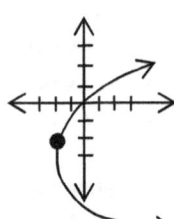

C)   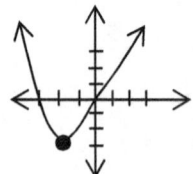   D)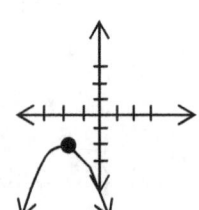

44. Find the zeroes of
$f(x) = x^3 + x^2 - 14x - 24$
(Rigorous)(Skill 0005)

A) 4, 3, 2
B) 3, -8
C) 7, -2, -1
D) 4, -3, -2

45. Which equation corresponds to the logarithmic statement: $\log_x k = m$?
(Rigorous)(Skill 0005)

A) $x^m = k$
B) $k^m = x$
C) $x^k = m$
D) $m^x = k$

46. Solve for x $10^{x-3} + 5 = 105$.
(Rigorous)(Skill 0005)

A) 3
B) 10
C) 2
D) 5

47. L'Hospital's rule provides a method to evaluate which of the following?
(Easy)(Skill 0006)

A) Limit of a function
B) Derivative of a function
C) Sum of an arithmetic series
D) Sum of a geometric series

48. Find the area under the function $y = x^2 + 4$ from $x = 3$ to $x = 6$.
(Average Rigor)(Skill 0006)

A) 75
B) 21
C) 96
D) 57

49. If the velocity of a body is given by v = 16 - t², find the distance traveled from t = 0 until the body comes to a complete stop.
(Average Rigor)(Skill 0006)

A) 16
B) 43
C) 48
D) 64

50. Find the following limit:
$\lim_{x \to 2} \dfrac{x^2 - 4}{x - 2}$
(Average Rigor)(Skill 0006)

A) 0
B) Infinity
C) 2
D) 4

51. Find the first derivative of the function:
$f(x) = x^3 - 6x^2 + 5x + 4$
(Rigorous)(Skill 0006)

A) $3x^3 - 12x^2 + 5x = f'(x)$
B) $3x^2 - 12x - 5 = f'(x)$
C) $3x^2 - 12x + 9 = f'(x)$
D) $3x^2 - 12x + 5 = f'(x)$

52. Find the absolute maximum obtained by the function $y = 2x^2 + 3x$ on the interval $x = 0$ to $x = 3$.
    (Rigorous)(Skill 0006)

    A) −3/4
    B) −4/3
    C) 0
    D) 27

53. Find the antiderivative for $4x^3 - 2x + 6 = y$.
    (Rigorous)(Skill 0006)

    A) $x^4 - x^2 + 6x + C$
    B) $x^4 - 2/3x^3 + 6x + C$
    C) $12x^2 - 2 + C$
    D) $4/3x^4 - x^2 + 6x + C$

54. Find the antiderivative for the function $y = e^{3x}$.
    (Rigorous)(Skill 0006)

    A) $3x(e^{3x}) + C$
    B) $3(e^{3x}) + C$
    C) $1/3(e^x) + C$
    D) $1/3(e^{3x}) + C$

55. Evaluate $\int_0^2 (x^2 + x - 1) dx$
    (Rigorous)(Skill 0006)

    A) 11/3
    B) 8/3
    C) -8/3
    D) -11/3

56. Find the following limit:
    $\lim_{x \to 0} \frac{\sin 2x}{5x}$
    (Rigorous)(Skill 0006)

    A) Infinity
    B) 0
    C) 1.4
    D) 1

57. Compute the standard deviation for the following set of temperatures.
    (37, 38, 35, 37, 38, 40, 36, 39)
    (Easy)(Skill 0007)

    A) 37.5
    B) 1.5
    C) 0.5
    D) 2.5

58. Compute the median for the following data set:
    {12, 19, 13, 16, 17, 14}
    (Easy)(Skill 0007)

    A) 14.5
    B) 15.17
    C) 15
    D) 16

59. A jar contains 3 red marbles, 5 white marbles, 1 green marble and 15 blue marbles. If one marble is picked at random from the jar, what is the probability that it will be red?
    (Easy)(Skill 0007)

    A) 1/3
    B) 1/8
    C) 3/8
    D) 1/24

60. Which of the following is not a valid method of collecting statistical data? (Average Rigor)(Skill 0007)

    A) Random sampling
    B) Systematic sampling
    C) Cluster sampling
    D) Cylindrical sampling

61. The probability distribution shown below exhibits: (Average Rigor)(Skill 0007)

    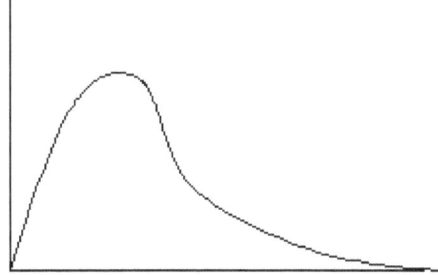

    A) positive skew
    B) negative skew
    C) excessive kurtosis
    D) diminished kurtosis

62. Half the students in a class scored 80% on an exam, most of the rest scored 85% except for one student who scored 10%. Which would be the best measure of central tendency for the test scores? (Rigorous)(Skill 0007)

    A) mean
    B) median
    C) mode
    D) either the median or the mode because they are equal

63. If the correlation between two variables is given as zero, the association between the two variables is (Rigorous)(Skill 0007)

    A) negative linear
    B) positive linear
    C) quadratic
    D) random

64. A die is rolled several times. What is the probability that a 3 will not appear before the third roll of the die? (Rigorous)(Skill 0007)

    A) 1/3
    B) 25/216
    C) 25/36
    D) 1/216

65. If there are three people in a room, what is the probability that at least two of them will share a birthday? (Assume a year has 365 days) (Rigorous)(Skill 0007)

    A) 0.67
    B) 0.05
    C) 0.008
    D) 0.33

66. The number of pizza slices eaten per college student per year fits a normal distribution with a mean of 55 and a standard deviation of 15. The number of pizza slices eaten annually by the students in the top 2.5% of the distribution is greater than:
(Rigorous)(Skill 0007)

A) 70
B) 85
C) 100
D) 110

67. The scalar multiplication of the number 3 with the matrix $\begin{pmatrix} 2 & 1 \\ 3 & 5 \end{pmatrix}$ yields
(Easy)(Skill 0008)

A) 33
B) $\begin{pmatrix} 6 & 1 \\ 9 & 5 \end{pmatrix}$
C) $\begin{pmatrix} 2 & 3 \\ 3 & 15 \end{pmatrix}$
D) $\begin{pmatrix} 6 & 3 \\ 9 & 15 \end{pmatrix}$

68. The result of adding the following matrices is
$\begin{pmatrix} 6 & 3 \\ 9 & 15 \end{pmatrix} + \begin{pmatrix} 4 & 7 \\ 1 & 0 \end{pmatrix}$
(Easy)(Skill 0008)

A) $\begin{pmatrix} 10 & 10 \\ 10 & 15 \end{pmatrix}$
B) $\begin{pmatrix} 13 & 7 \\ 9 & 16 \end{pmatrix}$
C) 45
D) $\begin{pmatrix} 20 \\ 25 \end{pmatrix}$

69. The product of two matrices can be found only if
(Easy)(Skill 0008)

A) The number of rows in the first matrix is equal to the number of rows in the second matrix
B) The number of columns in the first matrix is equal to the number of columns in the second matrix
C) The number of columns in the first matrix is equal to the number of rows in the second matrix
D) The number of rows in the first matrix is equal to the number of columns in the second matrix

70. Solve the following matrix equation

$$3x + \begin{pmatrix} 1 & 5 & 2 \\ 0 & 6 & 9 \end{pmatrix} = \begin{pmatrix} 7 & 17 & 5 \\ 3 & 9 & 9 \end{pmatrix}$$

(Average Rigor)(Skill 0008)

A) $\begin{pmatrix} 2 & 4 & 1 \\ 1 & 1 & 0 \end{pmatrix}$

B) 2

C) $\begin{pmatrix} 8 & 23 & 7 \\ 3 & 15 & 18 \end{pmatrix}$

D) $\begin{pmatrix} 9 \\ 2 \end{pmatrix}$

71. Find the value of the determinant of the matrix. (Average Rigor)(Skill 0008)

$$\begin{vmatrix} 2 & 1 & -1 \\ 4 & -1 & 4 \\ 0 & -3 & 2 \end{vmatrix}$$

A) 0
B) 23
C) 24
D) 40

72. Evaluate the following matrix product:

$$\begin{pmatrix} 2 & 1 & 3 \\ 2 & 2 & 4 \end{pmatrix} \times \begin{pmatrix} 6 & 5 \\ 2 & 1 \\ 2 & 7 \end{pmatrix}$$

(Rigorous)(Skill 0008)

A) $\begin{pmatrix} 20 & 32 & 24 \\ 24 & 40 & 48 \end{pmatrix}$

B) $\begin{pmatrix} 20 & 32 \\ 40 & 24 \\ 24 & 48 \end{pmatrix}$

C) 116

D) $\begin{pmatrix} 20 & 32 \\ 24 & 40 \end{pmatrix}$

73. What conclusion can be drawn from the graph below?

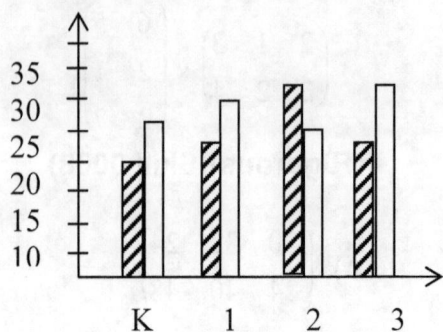

MLK Elementary Student Enrollment
☒ Boys ☐ Girls
(Easy)(Skill 0009)

A) The number of students in first grade exceeds the number in second grade.
B) There are more boys than girls in the entire school.
C) There are more girls than boys in the first grade.
D) Third grade has the largest number of students.

74. The pie chart below shows sales at an automobile dealership for the first four months of a year. What percentage of the vehicles were sold in April? (Easy)(Skill 0009)

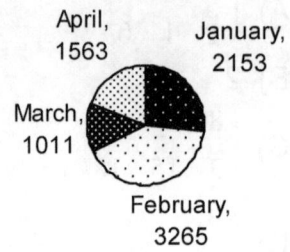

A) More than 50%
B) Less than 25%
C) Between 25% and 50%
D) None

75. Which of the following types of graphs would be best to use to record the eye color of the students in the class?
(Average Rigor)(Skill 0009)

A) Bar graph or circle graph
B) Pictograph or bar graph
C) Line graph or pictograph
D) Line graph or bar graph

76. Which type of graph uses symbols to represent quantities?
    (Average Rigor)(Skill 0009)

    A) Bar graph
    B) Line graph
    C) Pictograph
    D) Circle graph

77. Which of the following sets is closed under division?
    (Average Rigor)(Skill 0009)

    I) {½, 1, 2, 4}
    II) {-1, 1}
    III) {-1, 0, 1}

    A) I only
    B) II only
    C) III only
    D) I and II

78. Determine the number of subsets of set K.
    $K = \{4, 5, 6, 7\}$
    (Average Rigor)(Skill 0009)

    A) 15
    B) 16
    C) 17
    D) 18

79. What is the sum of the first 20 terms of the geometric sequence (2,4,8,16,32,...)?
    (Average Rigor)(Skill 0009)

    A) 2097150
    B) 1048575
    C) 524288
    D) 1048576

80. Find the sum of the first one hundred terms in the progression.
    (-6, -2, 2 . . .)
    (Rigorous)(Skill 0009)

    A) 19,200
    B) 19,400
    C) -604
    D) 604

TEACHER CERTIFICATION STUDY GUIDE

Constructed Response Questions

Problem

Is $y = 3x - 6$ a bisector of the line segment with endpoints at (2, 4) and (8, -1)?

Model

The town of Verdant Slopes has been experiencing a boom in population growth. By the year 2000, the population had grown to 45,000, and by 2005, the population had reached 60,000.

a. Using the formula for slope as a model, find the average rate of change in population growth, expressing your answer in people per year.

b. Using the average rate of change determined in a., predict the population of Verdant Slopes in the year 2010.

Proof

Prove that $\cot x + \tan x = (\csc x)(\sec x)$.

Answer Key

1.	B	23.	D	45.	A	67.	D
2.	C	24.	C	46.	D	68.	A
3.	A	25.	A	47.	A	69.	C
4.	C	26.	B	48.	A	70.	A
5.	D	27.	C	49.	B	71.	C
6.	D	28.	B	50.	D	72.	D
7.	A	29.	C	51.	D	73.	B
8.	A	30.	A	52.	D	74.	B
9.	D	31.	B	53.	A	75.	B
10.	C	32.	D	54.	D	76.	C
11.	C	33.	D	55.	B	77.	B
12.	A	34.	C	56.	C	78.	B
13.	A	35.	D	57.	B	79.	A
14.	C	36.	B	58.	C	80.	A
15.	C	37.	B	59.	B		
16.	B	38.	D	60.	D		
17.	B	39.	C	61.	A		
18.	A	40.	D	62.	B		
19.	B	41.	B	63.	D		
20.	C	42.	B	64.	B		
21.	C	43.	C	65.	C		
22.	A	44.	D	66.	B		

Rigor Table

Questions	Assessments
1	Easy
2	Average Rigor
3	Average Rigor
4	Average Rigor
5	Average Rigor
6	Average Rigor
7	Rigorous
8	Rigorous
9	Rigorous
10	Rigorous
11	Easy
12	Average Rigor
13	Average Rigor
14	Average Rigor
15	Rigorous
16	Rigorous
17	Rigorous
18	Rigorous
19	Rigorous
20	Easy
21	Average Rigor
22	Average Rigor
23	Average Rigor
24	Average Rigor
25	Rigorous
26	Rigorous
27	Rigorous
28	Rigorous
29	Rigorous
30	Easy
31	Easy
32	Average Rigor
33	Average Rigor
34	Average Rigor
35	Rigorous
36	Rigorous
37	Easy
38	Average Rigor
39	Average Rigor
40	Average Rigor
41	Average Rigor
42	Average Rigor
43	Average Rigor

44	Rigorous
45	Rigorous
46	Rigorous
47	Easy
48	Average Rigor
49	Average Rigor
50	Average Rigor
51	Rigorous
52	Rigorous
53	Rigorous
54	Rigorous
55	Rigorous
56	Rigorous
57	Easy
58	Easy
59	Easy
60	Average Rigor
61	Average Rigor
62	Rigorous
63	Rigorous
64	Rigorous
65	Rigorous
66	Rigorous
67	Easy
68	Easy
69	Easy
70	Average Rigor
71	Average Rigor
72	Rigorous
73	Easy
74	Easy
75	Average Rigor
76	Average Rigor
77	Average Rigor
78	Average Rigor
79	Average Rigor
80	Rigorous
Easy 19%	1,11,20,30,31,37,47,57,58,59,67,68,69,73,74
Average Rigor 41%	2,3,4,5,6,12,13,14,21,22,23,24,32,33,34,38,39,40,41,42, 43,48,49,50,60,61,70,71,75,76,77,78,79
Rigorous 40%	7,8,9,10,15,16,17,18,19,25,26,27,28,29,35,36,44,45,46,51, 52,53,54,55,56,62,63,64,65,66,72,80

TEACHER CERTIFICATION STUDY GUIDE

Rationales with Sample Questions

1. Find the LCM of 27, 90 and 84.
   (Easy)(Skill 0001)

   A) 90
   B) 3780
   C) 204120
   D) 1260

Answer: B

To find the LCM of the above numbers, factor each into its prime factors and multiply each common factor the maximum number of times it occurs. Thus 27=3x3x3; 90=2x3x3x5; 84=2x2x3x7; LCM = 2x2x3x3x3x5x7=3780.

2. Solve for x by factoring $2x^2 - 3x - 2 = 0$.
   (Average Rigor) (Skill 0001)

   A) x = (-1,2)
   B) x = (0.5,-2)
   C) x=(-0.5,2)
   D) x=(1,-2)

Answer: C

$2x^2 - 3x - 2 = 2x^2 - 4x + x - 2 = 2x(x-2) + (x-2) = (2x+1)(x-2) = 0.$
Thus x = -0.5 or 2.

3. What would be the total cost of a suit for $295.99 and a pair of shoes for $69.95 including 6.5% sales tax?
   (Average Rigor) (Skill 0001)

   A) $389.73
   B) $398.37
   C) $237.86
   D) $315.23

Answer: A

Before the tax, the total comes to $365.94. Then .065(365.94) = 23.79. With the tax added on, the total bill is 365.94 + 23.79 = $389.73. (Quicker way: 1.065(365.94) = 389.73.)

MATHEMATICS

TEACHER CERTIFICATION STUDY GUIDE

4. Which of the following is always composite if *x* is odd, *y* is even, and both *x* and *y* are greater than or equal to 2?
   (Average Rigor)(Skill 0001)

   A) $x+y$
   B) $3x+2y$
   C) $5xy$
   D) $5x+3y$

Answer: C

A composite number is a number which is not prime. The prime number sequence begins 2,3,5,7,11,13,17,…. To determine which of the expressions is <u>always</u> composite, experiment with different values of x and y, such as x=3 and y=2, or x=5 and y=2. It turns out that 5xy will always be an even number, and therefore, composite, if y=2.

5. What is the smallest number that is divisible by 3 and 5 and leaves a remainder of 3 when divided by 7?
   (Average Rigor)(Skill 0001)

   A) 15
   B) 18
   C) 25
   D) 45

Answer: D

To be divisible by both 3 and 5, the number must be divisible by 15. Inspecting the first few multiples of 15, you will find that 45 is the first of the sequence that is 4 greater than a multiple of 7.

TEACHER CERTIFICATION STUDY GUIDE

6. Given the series of examples below, what is 5¢4?
   (Average Rigor)(Skill 0001)

   4¢3=13 7¢2=47
   3¢1=8 1¢5=-4

   A) 20
   B) 29
   C) 1
   D) 21

Answer: D

By observation of the examples given, $a \not{c} b = a^2 - b$. Therefore, $5 \not{c} 4 = 25 - 4 = 21$.

7. Which of the following is a factor of the expression $9x^2 + 6x - 35$?
   (Rigorous)(Skill 0001)

   A) 3x-5
   B) 3x-7
   C) x+3
   D) x-2

Answer: A

Recognize that the given expression can be written as the sum of two squares and utilize the formula $a^2 - b^2 = (a+b)(a-b)$.
$9x^2 + 6x - 35 = (3x+1)^2 - 36 = (3x+1+6)(3x+1-6) = (3x+7)(3x-5)$.

MATHEMATICS

8. **Which of the following is a factor of** $6 + 48m^3$
   (Rigorous) (Skill 0001)

   A) $(1 + 2m)$
   B) $(1 - 8m)$
   C) $(1 + m - 2m)$
   D) $(1 - m + 2m)$

Answer: A

Removing the common factor of 6 and then factoring the sum of two cubes gives $6 + 48m^3 = 6(1 + 8m^3) = 6(1 + 2m)(1^2 - 2m + (2m)^2)$.

9. **Which of the following is incorrect?**
   (Rigorous) (Skill 0001)

   A) $(x^2 y^3)^2 = x^4 y^6$
   B) $m^2 (2n)^3 = 8m^2 n^3$
   C) $(m^3 n^4)/(m^2 n^2) = mn^2$
   D) $(x + y^2)^2 = x^2 + y^4$

Answer: D

Using FOIL to do the expansion, we get $(x + y^2)^2 = (x + y^2)(x + y^2) = x^2 + 2xy^2 + y^4$.

10. **Solve for** x: $18 = 4 + |2x|$
    (Rigorous) (Skill 0001)

    A) $\{-11, 7\}$
    B) $\{-7, 0, 7\}$
    C) $\{-7, 7\}$
    D) $\{-11, 11\}$

Answer: C

Using the definition of absolute value, two equations are possible: $18 = 4 + 2x$ or $18 = 4 - 2x$. Solving for x gives x = 7 or x = -7.

TEACHER CERTIFICATION STUDY GUIDE

11. Find the surface area of a box which is 3 feet wide, 5 feet tall, and 4 feet deep.
    (Easy)(Skill 0002)

    A) 47 sq. ft.
    B) 60 sq. ft.
    C) 94 sq. ft
    D) 188 sq. ft.

Answer: C

Let's assume the base of the rectangular solid (box) is 3 by 4, and the height is 5. Then the surface area of the top and bottom together is 2(12) = 24. The sum of the areas of the front and back are 2(15) = 30, while the sum of the areas of the sides are 2(20)=40. The total surface area is therefore 94 square feet.

12. The term "cubic feet" indicates which kind of measurement?
    (Average Rigor)(Skill 0002)

    A) Volume
    B) Mass
    C) Length
    D) Distance

Answer: A

The word *cubic* indicates that this is a term describing volume.

13. Given a 30 meter x 60 meter garden with a circular fountain with a 5 meter radius, calculate the area of the portion of the garden not occupied by the fountain.
    (Average Rigor)(Skill 0002)

    A) 1721 m²
    B) 1879 m²
    C) 2585 m²
    D) 1015 m²

Answer: A

Find the area of the garden and then subtract the area of the fountain:
$30(60) - \pi(5)^2$ or approximately 1721 square meters.

14. **Find the height of a box with surface area of 94 sq. ft. with a width of 3 feet and a depth of 4 feet.**
    (Average Rigor)(Skill 0002)

    A) 3 ft.
    B) 4 ft.
    C) 5 ft
    D) 6 ft.

Answer: C

94 = 2(3h) + 2(4h) + 2(12)
94 = 6h + 8h + 24
94 = 14h + 24
70 = 14h
5 = h

15. **If the area of the base of a cone is tripled, the volume will be (Rigorous)(Skill 0002)**

    A) the same as the original
    B) 9 times the original
    C) 3 times the original
    D) 3π times the original

Answer: C

The formula for the volume of a cone is $V = \frac{1}{3}Bh$, where B is the area of the circular base and h is the height. If the area of the base is tripled, the volume becomes
$V = \frac{1}{3}(3B)h = Bh$, or three times the original area.

16. Find the area of the figure pictured below.
    (Rigorous)(Skill 0002)

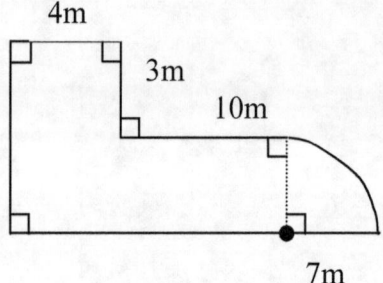

A) 136.47 m²
B) 148.48 m²
C) 293.86 m²
D) 178.47 m²

Answer: B

Divide the figure into 2 rectangles and one quarter circle. The tall rectangle on the left will have dimensions 10 by 4 and area 40. The rectangle in the center will have dimensions 7 by 10 and area 70. The quarter circle will have area $.25(\pi)7^2 = 38.48$.
The total area is therefore approximately 148.48.

17. Compute the area of the shaded region, given a radius of 5 meters. O is the center.
    (Rigorous)(Skill 0002)

A) 7.13 cm²
B) 7.13 m²
C) 78.5 m²
D) 19.63 m²

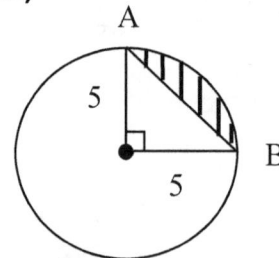

Answer: B

Area of triangle AOB is $.5(5)(5) = 12.5$ square meters. Since $\frac{90}{360} = .25$, the area of sector AOB (pie-shaped piece) is approximately $.25(\pi)5^2 = 19.63$. Subtracting the triangle area from the sector area to get the area of segment AB, we get approximately $19.63 - 12.5 = 7.13$ square meters.

18. **The length of a picture frame is 2 inches greater than its width. If the area of the frame is 143 square inches, what is its width?**
    (Rigorous)(Skill 0002)

    A) 11 inches
    B) 13 inches
    C) 12 inches
    D) 10 inches

Answer: A

First set up the equation for the problem. If the width of the picture frame is w, then w(w+2) = 143. Next, solve the equation to obtain w. Using the method of completing squares we have: $w^2 + 2w + 1 = 144; (w+1)^2 = 144; w + 1 = \pm 12$. Thus w = 11 or -13. Since the width cannot be negative, the correct answer is 11.

19. **Determine the area of the shaded region of the trapezoid in terms of *x* and *y*.**
    (Rigorous)(Skill 0002)

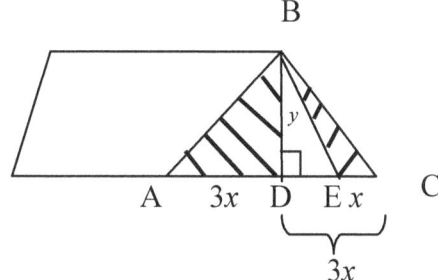

    A) $4xy$
    B) $2xy$
    C) $3x^2 y$
    D) There is not enough information given.

Answer: B

To find the area of the shaded region, find the area of triangle ABC and then subtract the area of triangle DBE. The area of triangle ABC is .5(6x)(y) = 3xy. The area of triangle DBE is .5(2x)(y) = xy. The difference is 2xy.

20. When you begin by assuming the conclusion of a theorem is false, then show that through a sequence of logically correct steps you contradict an accepted fact, this is known as
(Easy)(Skill 0003)

   A) inductive reasoning
   B) direct proof
   C) indirect proof
   D) exhaustive proof

Answer: C

By definition this describes the procedure of an indirect proof.

21. Which theorem can be used to prove $\triangle BAK \cong \triangle MKA$?
(Average Rigor)(Skill 0003)

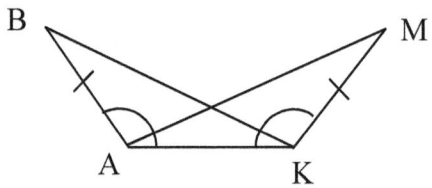

   A) SSS
   B) ASA
   C) SAS
   D) AAS

Answer: C

Since side AK is common to both triangles, the triangles can be proved congruent by using the Side-Angle-Side Postulate.

TEACHER CERTIFICATION STUDY GUIDE

22. **Choose the correct statement concerning the median and altitude in a triangle.**
    (Average Rigor)(Skill 0003)

    A) The median and altitude of a triangle may be the same segment.
    B) The median and altitude of a triangle are always different segments.
    C) The median and altitude of a right triangle are always the same segment.
    D) The median and altitude of an isosceles triangle are always the same segment.

Answer: A

The most one can say with certainty is that the median (segment drawn to the midpoint of the opposite side) and the altitude (segment drawn perpendicular to the opposite side) of a triangle <u>may</u> coincide, but they more often do not. In an isosceles triangle, the median and the altitude to the <u>base</u> are the same segment.

23. **Compute the distance from (-2, 7) to the line $x = 5$.**
    (Average Rigor)(Skill 0003)

    A) -9
    B) -7
    C) 5
    D) 7

Answer: D

The line $x = 5$ is a vertical line passing through (5,0) on the Cartesian plane. By observation the distance along the horizontal line from the point (-2,7) to the line x=5 is 7 units.

MATHEMATICS

24. Which of the following statements about a trapezoid is incorrect?
(Average Rigor)(Skill 0003)

A) It has one pair of parallel sides
B) The parallel sides are called bases
C) If the two bases are the same length, the trapezoid is called isosceles
D) The median is parallel to the bases

Answer: C

A trapezoid is isosceles if the two legs (not bases) are the same length.

25. Given $K(-4, y)$ and $M(2,-3)$ with midpoint $L(x,1)$, determine the values of x and y.
(Rigorous)(Skill 0003)

A) $x = -1, y = 5$
B) $x = 3, y = 2$
C) $x = 5, y = -1$
D) $x = -1, y = -1$

Answer: A

The formula for finding the midpoint (a,b) of a segment passing through the points (x_1, y_1) and (x_2, y_2) is $(a,b) = (\frac{x_1 + x_2}{2}, \frac{y_1 + y_2}{2})$. Setting up the corresponding equations from this information gives us $x = \frac{-4 + 2}{2}$, and $1 = \frac{y - 3}{2}$. Solving for x and y gives x = -1 and y = 5.

26. Which equation represents a circle with a diameter whose endpoints are $(0,7)$ and $(0,3)$?
   (Rigorous)(Skill 0003)

   A) $x^2 + y^2 + 21 = 0$
   B) $x^2 + y^2 - 10y + 21 = 0$
   C) $x^2 + y^2 - 10y + 9 = 0$
   D) $x^2 - y^2 - 10y + 9 = 0$

Answer: B

With a diameter going from (0,7) to (0,3), the diameter of the circle must be 4, the radius must be 2, and the center of the circle must be at (0,5). Using the standard form for the equation of a circle, we get $(x-0)^2 + (y-5)^2 = 2^2$. Expanding, we get $x^2 + y^2 - 10y + 21 = 0$.

27. What is the degree measure of an interior angle of a regular 10 sided polygon?
   (Rigorous)(Skill 0003)

   A) 18°
   B) 36°
   C) 144°
   D) 54°

Answer: C

Formula for finding the measure of each interior angle of a regular polygon with n sides is $\frac{(n-2)180}{n}$. For n=10, we get $\frac{8(180)}{10} = 144$.

28. **What is the measure of minor arc AD, given measure of arc PS is 40° and $m\angle K = 10°$?**
    (Rigorous)(Skill 0003)

    A) 50°
    B) 20°
    C) 30°
    D) 25°

Answer: B

The formula relating the measure of angle K and the two arcs it intercepts is $m\angle K = \frac{1}{2}(mPS - mAD)$. Substituting the known values, we get $10 = \frac{1}{2}(40 - mAD)$. Solving for mAD gives an answer of 20 degrees.

29. **Find the length of the major axis of $x^2 + 9y^2 = 36$.**
    (Rigorous)(Skill 0003)

    A) 4
    B) 6
    C) 12
    D) 8

Answer: C

Dividing by 36, we get $\frac{x^2}{36} + \frac{y^2}{4} = 1$, which tells us that the ellipse intersects the x-axis at 6 and –6, and therefore the length of the major axis is 12. (The ellipse intersects the y-axis at 2 and –2).

30. **The cosine function is defined as**
    (Easy)(Skill 0004)

    A) base/hypotenuse
    B) perpendicular/hypotenuse
    C) hypotenuse/base
    D) perpendicular/base

Answer: A

The cosine of an angle in a right triangle is defined as the ratio of the base (the side adjacent to the angle) to the hypotenuse (side opposite the right angle).

31. **The following is a Pythagorean identity:**
    (Easy)(Skill 0004)

    A) $\sin^2\theta - \cos^2\theta = 1$
    B) $\sin^2\theta + \cos^2\theta = 1$
    C) $\cos^2\theta - \sin^2\theta = 1$
    D) $\cos^2\theta + \tan^2\theta = 1$

Answer: B

The Pythagorean identity $\sin^2\theta + \cos^2\theta = 1$ is derived from the definitions of the sine and cosine functions and Pythagorean Theorem of geometry.

32. **Determine the measures of angles A and B.**
    (Average Rigor)(Skill 0004)

    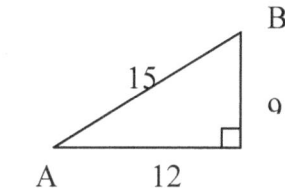

    A) A = 30°, B = 60°
    B) A = 60°, B = 30°
    C) A = 53°, B = 37°
    D) A = 37°, B = 53°

Answer: D

Tan A = 9/12=.75 and tan^{-1}.75 = 37 degrees. Since angle B is complementary to angle A, the measure of angle B is therefore 53 degrees.

33. **Which expression is not identical to sinx?**

    (Average Rigor)(Skill 0004)

    A) $\sqrt{1-\cos^2 x}$
    B) $\tan x \cos x$
    C) $1/\csc x$
    D) $1/\sec x$

Answer: D

Using the basic definitions of the trigonometric functions and the Pythagorean identity, we see that the first three options are all identical to sinx. secx= 1/cosx is not the same as sinx.

TEACHER CERTIFICATION STUDY GUIDE

34. Determine the rectangular coordinates of the point with polar coordinates (5, 60°).
    (Average Rigor)(Skill 0004)

    A) (0.5, 0.87)
    B) (-0.5, 0.87)
    C) (2.5, 4.33)
    D) (25, 150°)

Answer: C

Given the polar point $(r, \theta) = (5, 60)$, we can find the rectangular coordinates this way: $(x,y) = (r\cos\theta, r\sin\theta) = (5\cos 60, 5\sin 60) = (2.5, 4.33)$.

35. Which expression is equivalent to $1 - \sin^2 x$?
    (Rigorous)(Skill 0004)

    A) $1 - \cos^2 x$
    B) $1 + \cos^2 x$
    C) $1/\sec x$
    D) $1/\sec^2 x$

Answer: D

Using the Pythagorean Identity, we know $\sin^2 x + \cos^2 x = 1$. Thus $1 - \sin^2 x = \cos^2 x$, which by definition is equal to $1/\sec^2 x$.

36. For an acute angle x, sinx = 3/5. What is cotx?
    (Rigorous)(Skill 0004)

    A) 5/3
    B) 3/4
    C) 1.33
    D) 1

Answer: B

Using the Pythagorean Identity, we know $\sin^2 x + \cos^2 x = 1$. Thus

$\cos x = \sqrt{1 - \dfrac{9}{25}} = \dfrac{4}{5}; \cot x = \dfrac{\cos x}{\sin x} = \dfrac{4}{3}$.

37. **What is the equation of the graph shown below?**
    (Easy)(Skill 0005)

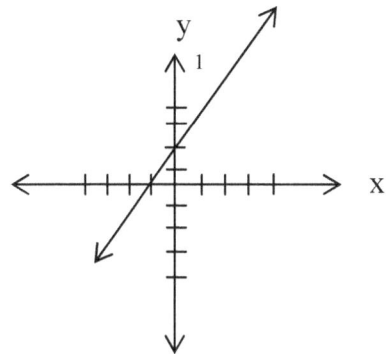

   A) $2x + y = 2$
   B) $2x - y = -2$
   C) $2x - y = 2$
   D) $2x + y = -2$

Answer: B

By observation, we see that the graph has a y-intercept of 2 and a slope of 2/1 = 2. Therefore its equation is y = mx + b = 2x + 2. Rearranging the terms gives 2x − y = -2.

38. **Solve for** v_0: $d = at(v_t - v_0)$
    (Average Rigor)(Skill 0005)

   A) $v_0 = atd - v_t$
   B) $v_0 = d - atv_t$
   C) $v_0 = atv_t - d$
   D) $v_0 = (atv_t - d)/at$

Answer: D

Using the Distributive Property and other properties of equality to isolate v_0 gives
d = atv$_t$ − atv$_0$,   atv$_0$ = atv$_t$ − d,   v$_0$ = $\dfrac{atv_t - d}{at}$.

39. Given $f(x) = 3x - 2$ and $g(x) = x^2$, determine $g(f(x))$.
    (Average Rigor)(Skill 0005)

    A) $3x^2 - 2$
    B) $9x^2 + 4$
    C) $9x^2 - 12x + 4$
    D) $3x^3 - 2$

Answer: C

The composite function g(f(x)) = (3x-2)² = 9x² – 12x + 4.

40. Which graph represents the solution set for $x^2 - 5x > -6$?
    (Average Rigor)(Skill 0005)

    A) ←―○―――○―→
         -2   0   2

    B) ←○―――――○―→
        -3   0   3

    C) ←―○―――○―――→
         -2   0   2

    D) ←――――○○―→
        -3   0 2 3

Answer: D

Rewriting the inequality gives x² – 5x + 6 > 0. Factoring gives (x – 2)(x – 3) > 0. The two cut-off points on the number line are now at x = 2 and x = 3. Choosing a random number in each of the three parts of the numberline, we test them to see if they produce a true statement. If x = 0 or x = 4, (x-2)(x-3)>0 is true. If x = 2.5, (x-2)(x-3)>0 is false. Therefore the solution set is all numbers smaller than 2 or greater than 3.

TEACHER CERTIFICATION STUDY GUIDE

41. **Solve for x:** $6(x+2) - 5 = 21 - x$.
    (Average Rigor)(Skill 0005)

    A) 3
    B) 2
    C) 5
    D) 0

Answer: B

$6(x+2) - 5 = 21 - x \Rightarrow 6x + 12 - 5 = 21 - x \Rightarrow 7x = 21 - 7 \Rightarrow 7x = 14 \Rightarrow x = 2$

42. **How does the function** $y = x^3 + x^2 + 4$ **behave from** $x = 1$ **to** $x = 3$?
    (Average Rigor) (Skill 0005)

    A) increasing, then decreasing
    B) increasing
    C) decreasing
    D) neither increasing nor decreasing

Answer: B

To find critical points, take the derivative, set it equal to 0, and solve for x. f'(x) = 3x² + 2x = x(3x+2)=0. CP at x=0 and x=-2/3. Neither of these CP is on the interval from x=1 to x=3. Testing the endpoints: at x=1, y=6 and at x=3, y=38. Since the derivative is positive for all values of x from x=1 to x=3, the curve is increasing on the entire interval.

TEACHER CERTIFICATION STUDY GUIDE

43. Which graph represents the equation of $y = x^2 + 3x$?
    (Average Rigor) (Skill 0005)

A)   B)

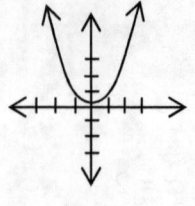   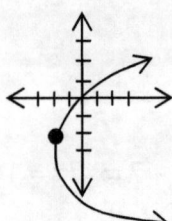

C)   D)

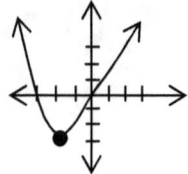   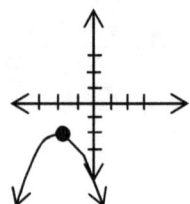

Answer: C

B is not the graph of a function. D is the graph of a parabola where the coefficient of x^2 is negative. A appears to be the graph of $y = x^2$. To find the x-intercepts of $y = x^2 + 3x$, set $y = 0$ and solve for x: $0 = x^2 + 3x = x(x + 3)$ to get $x = 0$ or $x = -3$. Therefore, the graph of the function intersects the x-axis at x=0 and x=-3.

44. Find the zeroes of $f(x) = x^3 + x^2 - 14x - 24$
    (Rigorous)(Skill 0005)

    A) 4, 3, 2
    B) 3, -8
    C) 7, -2, -1
    D) 4, -3, -2

Answer: D

Possible rational roots of the equation 0 = x^3 + x^2 – 14x -24 are all the positive and negative factors of 24. By substituting into the equation, we find that –2 is a root, and therefore that x+2 is a factor. By performing the long division (x^3 + x^2 – 14x – 24)/(x+2), we can find that another factor of the original equation is x^2 – x – 12 or (x-4)(x+3). Therefore the zeros of the original function are –2, -3, and 4.

45. Which equation corresponds to the logarithmic statement: $\log_x k = m$?
    (Rigorous)(Skill 0005)

    A) $x^m = k$
    B) $k^m = x$
    C) $x^k = m$
    D) $m^x = k$

Answer: A

By definition of log form and exponential form, $\log_x k = m$ corresponds to x^m = k.

46. **Solve for x** $10^{x-3} + 5 = 105$.
    (Rigorous)(Skill 0005)

    A) 3
    B) 10
    C) 2
    D) 5

Answer: D

$10^{x-3} = 100$. Taking the logarithm to base 10 of both sides $(x-3)\log_{10} 10 = \log_{10} 100; x - 3 = 2; x = 5$.

47. **L'Hospital's rule provides a method to evaluate which of the following?**
    (Easy)(Skill 0006)

    A) Limit of a function
    B) Derivative of a function
    C) Sum of an arithmetic series
    D) Sum of a geometric series

Answer: A

L'Hospital's rule is used to find the limit of a function by taking the derivatives of the numerator and denominator. Since the primary purpose of the rule is to find the limit, A is the correct answer.

48. Find the area under the function $y = x^2 + 4$ from $x = 3$ to $x = 6$.
    (Average Rigor)(Skill 0006)

    A) 75
    B) 21
    C) 96
    D) 57

Answer: A

To find the area set up the definite integral: $\int_{3}^{6}(x^2 + 4)dx = (\frac{x^3}{3} + 4x)$. Evaluate the expression at x=6, at x=3, and then subtract to get (72+24)−(9+12)=75.

49. If the velocity of a body is given by v = 16 - t², find the distance traveled from t = 0 until the body comes to a complete stop.
    (Average Rigor)(Skill 0006)

    A) 16
    B) 43
    C) 48
    D) 64

Answer: B

Recall that the derivative of the distance function is the velocity function. In reverse, the integral of the velocity function is the distance function. To find the time needed for the body to come to a stop when v=0, solve for t: $v = 16 - t^2 = 0$. Result: t = 4 seconds. The distance function is s = 16t - $\frac{t^3}{3}$. At t=4, s= 64 – 64/3 or approximately 43 units.

TEACHER CERTIFICATION STUDY GUIDE

50. Find the following limit: $\lim_{x \to 2} \dfrac{x^2 - 4}{x - 2}$

    (Average Rigor)(Skill 0006)

    A) 0
    B) Infinity
    C) 2
    D) 4

Answer: D

First factor the numerator and cancel the common factor to get the limit.

$$\lim_{x \to 2} \dfrac{x^2 - 4}{x - 2} = \lim_{x \to 2} \dfrac{(x-2)(x+2)}{(x-2)} = \lim_{x \to 2} (x+2) = 4$$

51. Find the first derivative of the function: $f(x) = x^3 - 6x^2 + 5x + 4$
    (Rigorous)(Skill 0006)

    A) $3x^3 - 12x^2 + 5x = f'(x)$
    B) $3x^2 - 12x - 5 = f'(x)$
    C) $3x^2 - 12x + 9 = f'(x)$
    D) $3x^2 - 12x + 5 = f'(x)$

Answer: D

Use the Power Rule for polynomial differentiation: if $y = ax^n$, then $y' = nax^{n-1}$.

MATHEMATICS

52. **Find the absolute maximum obtained by the function** $y = 2x^2 + 3x$ **on the interval** $x = 0$ **to** $x = 3$.
    (Rigorous)(Skill 0006)

    A) $-3/4$
    B) $-4/3$
    C) 0
    D) 27

Answer: D

Find CP at x=-.75 as done in #63. Since the CP is not in the interval from x=0 to x=3, just find the values of the functions at the endpoints. When x=0, y=0, and when x=3, y = 27. Therefore 27 is the absolute maximum on the given interval.

53. **Find the antiderivative for** $4x^3 - 2x + 6 = y$.
    (Rigorous)(Skill 0006)

    A) $x^4 - x^2 + 6x + C$
    B) $x^4 - 2/3x^3 + 6x + C$
    C) $12x^2 - 2 + C$
    D) $4/3x^4 - x^2 + 6x + C$

Answer: A

Use the rule for polynomial integration: given ax^n, the antiderivative is $\dfrac{ax^{n+1}}{n+1}$.

54. Find the antiderivative for the function $y = e^{3x}$.
    (Rigorous)(Skill 0006)

    A) $3x(e^{3x}) + C$
    B) $3(e^{3x}) + C$
    C) $1/3(e^x) + C$
    D) $1/3(e^{3x}) + C$

Answer: D

Use the rule for integration of functions of e: $\int e^x dx = e^x + C$.

55. Evaluate $\int_0^2 (x^2 + x - 1) dx$
    (Rigorous)(Skill 0006)

    A) 11/3
    B) 8/3
    C) -8/3
    D) -11/3

Answer: B

Use the fundamental theorem of calculus to find the definite integral: given a continuous function f on an interval [a,b], then $\int_a^b f(x)dx = F(b) - F(a)$, where F is an antiderivative of f.

$\int_0^2 (x^2 + x - 1) dx = (\frac{x^3}{3} + \frac{x^2}{2} - x)$ Evaluate the expression at x=2, at x=0, and then subtract to get 8/3 + 4/2 − 2−0 = 8/3.

56. Find the following limit: $\lim_{x \to 0} \dfrac{\sin 2x}{5x}$

    (Rigorous)(Skill 0006)

    A) Infinity
    B) 0
    C) 1.4
    D) 1

Answer: C

Since substituting x=0 will give an undefined answer, we can use L'Hospital's rule and take derivatives of both the numerator and denominator to find the limit.

$$\lim_{x \to 0} \frac{\sin 2x}{5x} = \lim_{x \to 0} \frac{2\cos 2x}{5} = \frac{2}{5} = 1.4$$

57. Compute the standard deviation for the following set of temperatures.
    (37, 38, 35, 37, 38, 40, 36, 39)
    (Easy)(Skill 0007)

    A) 37.5
    B) 1.5
    C) 0.5
    D) 2.5

Answer: B

Find the mean: 300/8 = 37.5. Then, using the formula for standard deviation, we get

$$\sqrt{\frac{2(37.5-37)^2 + 2(37.5-38)^2 + (37.5-35)^2 + (37.5-40)^2 + (37.5-36)^2 + (37.5-39)^2}{8}}$$

which has a value of 1.5.

TEACHER CERTIFICATION STUDY GUIDE

58. **Compute the median for the following data set:**
{12, 19, 13, 16, 17, 14}
(Easy)(Skill 0007)

A) 14.5
B) 15.17
C) 15
D) 16

Answer: C

Arrange the data in ascending order: 12,13,14,16,17,19. The median is the middle value in a list with an odd number of entries. When there is an even number of entries, the median is the mean of the two center entries. Here the average of 14 and 16 is 15.

59. **A jar contains 3 red marbles, 5 white marbles, 1 green marble and 15 blue marbles. If one marble is picked at random from the jar, what is the probability that it will be red?**
(Easy)(Skill 0007)

A) 1/3
B) 1/8
C) 3/8
D) 1/24

Answer: B

The total number of marbles is 24 and the number of red marbles is 3. Thus the probability of picking a red marble from the jar is 3/24=1/8.

60. Which of the following is not a valid method of collecting statistical data?
    (Average Rigor)(Skill 0007)

    A) Random sampling
    B) Systematic sampling
    C) Cluster sampling
    D) Cylindrical sampling

Answer: D

There is no such method as cylindrical sampling.

61. The probability distribution shown below exhibits:
    (Average Rigor)(Skill 0007)

    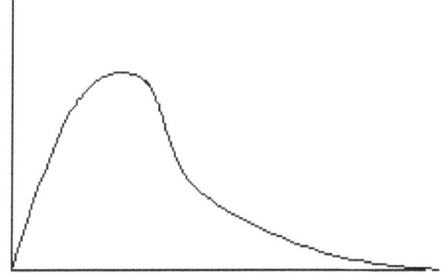

    A) positive skew
    B) negative skew
    C) excessive kurtosis
    D) diminished kurtosis

Answer: A

Positive skew denotes that distribution has an elongated **right** tail (compared to what is expected in a normal distribution).

62. Half the students in a class scored 80% on an exam, most of the rest scored 85% except for one student who scored 10%. Which would be the best measure of central tendency for the test scores?
    (Rigorous)(Skill 0007)

    A) mean
    B) median
    C) mode
    D) either the median or the mode because they are equal

Answer: B

In this set of data, the median (see #14) would be the most representative measure of central tendency since the median is independent of extreme values. Because of the 10% outlier, the mean (average) would be disproportionately skewed. In this data set, it is true that the median and the mode (number which occurs most often) are the same, but the median remains the best choice because of its special properties.

63. If the correlation between two variables is given as zero, the association between the two variables is
    (Rigorous)(Skill 0007)

    A) negative linear
    B) positive linear
    C) quadratic
    D) random

Answer: D

A correlation of 1 indicates a perfect positive linear association, a correlation of -1 indicates a perfect negative linear association while a correlation of zero indicates a random relationship between the variables.

64. **A die is rolled several times. What is the probability that a 3 will not appear before the third roll of the die?**
    (Rigorous)(Skill 0007)

    A) 1/3
    B) 25/216
    C) 25/36
    D) 1/216

Answer: B

The probability that a 3 will not appear before the third roll is the same as the probability that the first two rolls will consist of numbers other than 3. Since the probability of any one roll resulting in a number other than 3 is 5/6, the probability of the first two rolls resulting in a number other than 3 is (5/6) x (5/6) = 25/36.

65. **If there are three people in a room, what is the probability that at least two of them will share a birthday? (Assume a year has 365 days)**
    (Rigorous)(Skill 0007)

    A) 0.67
    B) 0.05
    C) 0.008
    D) 0.33

Answer: C

The best way to approach this problem is to use the fact that
the probability of an event + the probability of the event not happening = 1.
First find the probability that no two people will share a birthday and then subtract that from one.
The probability that two of the people will not share a birthday = 364/365 (since the second person's birthday can be one of the 364 days other than the birthday of the first person).
The probability that the third person will also not share either of the first two birthdays = (364/365) * (363/365) = 0.992.
Therefore, the probability that at least two people will share a birthday = 1 – 0.992 = 0.008.

66. The number of pizza slices eaten per college student per year fits a normal distribution with a mean of 55 and a standard deviation of 15. The number of pizza slices eaten annually by the students in the top 2.5% of the distribution is greater than:
(Rigorous)(Skill 0007)

A) 70
B) 85
C) 100
D) 110

Answer: B

Recall that in a normal distribution, **95%** of the observations fall within 2 standard deviations of the mean. That is, they fall between:

$\mu + 2\sigma$ and $\mu - 2\sigma$

Thus the top 2.5% will have a value greater than

$\mu + 2\sigma$

We can simply plug in the values provided in this problem to find:

$\mu + 2\sigma = 55 + 2 \times 15 = 85$

67. The scalar multiplication of the number 3 with the matrix $\begin{pmatrix} 2 & 1 \\ 3 & 5 \end{pmatrix}$ yields

(Easy)(Skill 0008)

A) 33
B) $\begin{pmatrix} 6 & 1 \\ 9 & 5 \end{pmatrix}$
C) $\begin{pmatrix} 2 & 3 \\ 3 & 15 \end{pmatrix}$
D) $\begin{pmatrix} 6 & 3 \\ 9 & 15 \end{pmatrix}$

Answer: D

In scalar multiplication of a matrix by a number, each element of the matrix is multiplied by that number.

TEACHER CERTIFICATION STUDY GUIDE

68. **The result of adding the following matrices is**

$$\begin{pmatrix} 6 & 3 \\ 9 & 15 \end{pmatrix} + \begin{pmatrix} 4 & 7 \\ 1 & 0 \end{pmatrix}$$

(Easy)(Skill 0008)

A) $\begin{pmatrix} 10 & 10 \\ 10 & 15 \end{pmatrix}$

B) $\begin{pmatrix} 13 & 7 \\ 9 & 16 \end{pmatrix}$

C) 45

D) $\begin{pmatrix} 20 \\ 25 \end{pmatrix}$

Answer: A

Two matrices with the same dimensions are added by adding the corresponding elements. In this case, element 1,1 (i.e. row 1, column 1) of matrix 1 was added to element 1,1 of matrix 2; element 2,1 of matrix 1 was added to element 2,1 of matrix 2; and so on for all four elements.

69. **The product of two matrices can be found only if**
    (Easy)(Skill 0008)

    A) The number of rows in the first matrix is equal to the number of rows in the second matrix
    B) The number of columns in the first matrix is equal to the number of columns in the second matrix
    C) The number of columns in the first matrix is equal to the number of rows in the second matrix
    D) The number of rows in the first matrix is equal to the number of columns in the second matrix

Answer: C

The number of columns in the first matrix must equal the number of rows in the second matrix since the process of multiplication involves multiplying the elements of every row of the first matrix with corresponding elements of every column of the second matrix.

70. Solve the following matrix equation

$$3x + \begin{pmatrix} 1 & 5 & 2 \\ 0 & 6 & 9 \end{pmatrix} = \begin{pmatrix} 7 & 17 & 5 \\ 3 & 9 & 9 \end{pmatrix}$$

(Average Rigor)(Skill 0008)

A) $\begin{pmatrix} 2 & 4 & 1 \\ 1 & 1 & 0 \end{pmatrix}$

B) 2

C) $\begin{pmatrix} 8 & 23 & 7 \\ 3 & 15 & 18 \end{pmatrix}$

D) $\begin{pmatrix} 9 \\ 2 \end{pmatrix}$

Answer: A

$$3x = \begin{pmatrix} 7 & 17 & 5 \\ 3 & 9 & 9 \end{pmatrix} - \begin{pmatrix} 1 & 5 & 2 \\ 0 & 6 & 9 \end{pmatrix} = \begin{pmatrix} 6 & 12 & 3 \\ 3 & 3 & 0 \end{pmatrix};$$

$$x = \frac{1}{3} \times \begin{pmatrix} 6 & 12 & 3 \\ 3 & 3 & 0 \end{pmatrix} = \begin{pmatrix} 2 & 4 & 1 \\ 1 & 1 & 0 \end{pmatrix}$$

71. Find the value of the determinant of the matrix.
(Average Rigor)(Skill 0008)

$$\begin{vmatrix} 2 & 1 & -1 \\ 4 & -1 & 4 \\ 0 & -3 & 2 \end{vmatrix}$$

A) 0
B) 23
C) 24
D) 40

Answer: C

To find the determinant of a matrix without the use of a graphing calculator, repeat the first two columns as shown,

```
2    1    -1   2    1
4    -1   4    4    -1
0    -3   2    0    -3
```

Starting with the top left-most entry, 2, multiply the three numbers in the diagonal going down to the right: 2(-1)(2)=-4. Do the same starting with 1: 1(4)(0)=0. And starting with −1: -1(4)(-3) = 12. Adding these three numbers, we get 8. Repeat the same process starting with the top right-most entry, 1. That is, multiply the three numbers in the diagonal going down to the left: 1(4)(2) = 8. Do the same starting with 2: 2(4)(-3) = -24 and starting with −1: -1(-1)(0) = 0. Add these together to get -16. To find the determinant, subtract the second result from the first: 8-(-16)=24.

TEACHER CERTIFICATION STUDY GUIDE

72. **Evaluate the following matrix product:**

$$\begin{pmatrix} 2 & 1 & 3 \\ 2 & 2 & 4 \end{pmatrix} \times \begin{pmatrix} 6 & 5 \\ 2 & 1 \\ 2 & 7 \end{pmatrix}$$

(Rigorous)(Skill 0008)

A) $\begin{pmatrix} 20 & 32 & 24 \\ 24 & 40 & 48 \end{pmatrix}$

B) $\begin{pmatrix} 20 & 32 \\ 40 & 24 \\ 24 & 48 \end{pmatrix}$

C) 116

D) $\begin{pmatrix} 20 & 32 \\ 24 & 40 \end{pmatrix}$

Answer: D

The product of a 2x3 matrix with a 3x2 matrix is a 2x2 matrix. This alone should be enough to identify the correct answer. Each term in the 2x2 matrix is calculated as described below.

Matrix 1, row 1 multiplied by matrix 2, column 1 gives entry 1, 1:
  2x6 + 1x2 + 3x2 = 12 + 2 + 6 = 20

Matrix 1, row 1 multiplied by matrix 2, column 2 gives entry 1, 2:
  2x5 + 1x1 + 3x7 = 10 + 1 + 21 = 32

Matrix 1, row 2 multiplied by matrix 2, column 1 gives entry 2, 1:
  2x6 + 2x2 + 4x2 = 12 + 4 + 8 = 24

Matrix 1, row 2 multiplied by matrix 2, column 2 gives entry 2, 2:
  2x5 + 2x1 + 4x7 = 10 + 2 + 28 = 40

73. What conclusion can be drawn from the graph below?

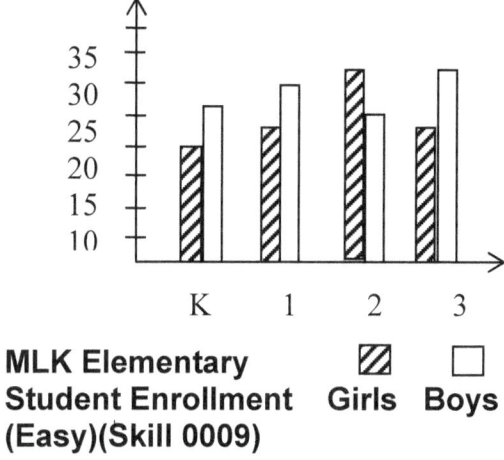

**MLK Elementary
Student Enrollment
(Easy)(Skill 0009)**

Girls Boys

A) The number of students in first grade exceeds the number in second grade.
B) There are more boys than girls in the entire school.
C) There are more girls than boys in the first grade.
D) Third grade has the largest number of students.

Answer: B

In Kindergarten, first grade, and third grade, there are more boys than girls. The number of extra girls in grade two is more than made up for by the extra boys in all the other grades put together.

TEACHER CERTIFICATION STUDY GUIDE

74. The pie chart below shows sales at an automobile dealership for the first four months of a year. What percentage of the vehicles were sold in April?
    (Easy)(Skill 0009)

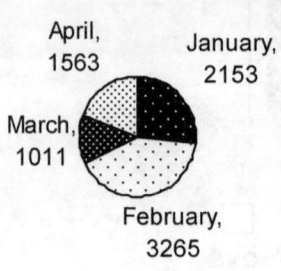

   A) More than 50%
   B) Less than 25%
   C) Between 25% and 50%
   D) None

Answer: B

It is clear from the chart that the April segment covers less than a quarter of the pie.

75. Which of the following types of graphs would be best to use to record the eye color of the students in the class?
    (Average Rigor)(Skill 0009)

   A) Bar graph or circle graph
   B) Pictograph or bar graph
   C) Line graph or pictograph
   D) Line graph or bar graph

Answer: B

A pictograph or a line graph could be used. In this activity, a line graph would not be used because it shows change over time. Although a circle graph could be used to show a percentage of students with brown eyes, blue eyes, etc. that representation would be too advanced for early childhood students.

TEACHER CERTIFICATION STUDY GUIDE

76. Which type of graph uses symbols to represent quantities? (Average Rigor)(Skill 0009)

- A) Bar graph
- B) Line graph
- C) Pictograph
- D) Circle graph

Answer: C

A pictograph shows comparison of quantities using symbols. Each symbol represents a number of items

77. Which of the following sets is closed under division? (Average Rigor)(Skill 0009)

$$\text{I) } \{½, 1, 2, 4\}$$
$$\text{II) } \{-1, 1\}$$
$$\text{III) } \{-1, 0, 1\}$$

- A) I only
- B) II only
- C) III only
- D) I and II

Answer: B

I is not closed because $\frac{4}{.5} = 8$ and 8 is not in the set.

III is not closed because $\frac{1}{0}$ is undefined.

II is closed because $\frac{-1}{1} = -1, \frac{1}{-1} = -1, \frac{1}{1} = 1, \frac{-1}{-1} = 1$ and all the answers are in the set.

MATHEMATICS

78. Determine the number of subsets of set K.
    K = {4, 5, 6, 7}
    (Average Rigor)(Skill 0009)

    A) 15
    B) 16
    C) 17
    D) 18

Answer: B

A set of n objects has 2^n subsets. Therefore, here we have $2^4 = 16$ subsets. These subsets include four which each have 1 element only, six which each have 2 elements, four which each have 3 elements, plus the original set, and the empty set.

79. What is the sum of the first 20 terms of the geometric sequence (2,4,8,16,32,…)?
    (Average Rigor)(Skill 0009)

    A) 2097150
    B) 1048575
    C) 524288
    D) 1048576

Answer: A

For a geometric sequence $a, ar, ar^2, ..., ar^n$, the sum of the first n terms is given by $\frac{a(r^n - 1)}{r - 1}$. In this case a=2 and r=2. Thus the sum of the first 20 terms of the sequence is given by $\frac{2(2^{20} - 1)}{2 - 1} = 2097150$.

80. **Find the sum of the first one hundred terms in the progression. (-6, -2, 2 . . .)**
   (Rigorous)(Skill 0009)

   A) 19,200
   B) 19,400
   C) -604
   D) 604

Answer: A

To find the 100th term: $t_{100} = -6 + 99(4) = 390$. To find the sum of the first 100 terms: $S = \dfrac{100}{2}(-6 + 390) = 19200$.

TEACHER CERTIFICATION STUDY GUIDE

Constructed Response Answers

Problem

Is $y = 3x - 6$ a bisector of the line segment with endpoints at (2, 4) and (8, -1)?

Find the midpoint of the line segment and then see if the midpoint is a point on the given line. Using the Midpoint Formula:

$$P = \left(\frac{2+8}{2}, \frac{4+(-1)}{2}\right) = \left(\frac{10}{2}, \frac{4-1}{2}\right) = \left(5, \frac{3}{2}\right) = (5, 1.5)$$

Check to see if this point is on the line:
$y = 3x - 6 = 3(5) - 6 = 15 - 6 = 9$

In order for this line to be a bisector, y must equal 1.5.
However, since $y = 9$, the answer to the question is "No, this is not a bisector."

Model

The town of Verdant Slopes has been experiencing a boom in population growth. By the year 2000, the population had grown to 45,000, and by 2005, the population had reached 60,000.

a. Using the formula for slope as a model, find the average rate of change in population growth, expressing your answer in people per year.

b. Using the average rate of change determined in a., predict the population of Verdant Slopes in the year 2010.

a. Let t represent the time and p represent population growth. The two observances are represented by (t_1, p_1) and (t_2, p_2).
1st observance = (t_1, p_1) = (2000, 45000)
2nd observance = (t_2, p_2) = (2005, 60000)

Use the formula for slope to find the average rate of change.

Rate of change = $\dfrac{p_2 - p_1}{t_2 - t_1}$

MATHEMATICS

Substitute values.

$$= \frac{60{,}000 - 45{,}000}{2005 - 2000}$$

Simplify.

$$= \frac{15{,}000}{5} = 3{,}000 \, people/year$$

The average rate of change in population growth for Verdant Slopes between the years 2000 and 2005 was 3,000 people/year.

b. 3,000 people/year x 5 years = 15,000 people
60,000 people + 15,000 people = 75,000 people

At a continuing average rate of growth of 3000 people/year, the population of Verdant Slopes could be expected to reach 75,000 by the year 2010.

Proof

Prove that $\cot x + \tan x = (\csc x)(\sec x)$.

$\cot x + \tan x =$

$\dfrac{\cos x}{\sin x} + \dfrac{\sin x}{\cos x}$ Reciprocal identities

$= \dfrac{\cos^2 x + \sin^2 x}{\sin x \cos x}$ Common denominator

$= \dfrac{1}{\sin x \cos x}$ Pythagorean identity

$= (\csc x)(\sec x)$ Reciprocal identity

Therefore,
$\cot x + \tan x = (\csc x)(\sec x)$

XAMonline, INC. 21 Orient Ave. Melrose, MA 02176

Toll Free number 800-509-4128

TO ORDER Fax 781-662-9268 OR www.XAMonline.com

WEST SERIES

PO#   Store/School:

Address 1:

Address 2 (Ship to other):

City, State Zip

Credit card number _____-_____-_____-_____   expiration _____

EMAIL _____

PHONE   FAX

ISBN	TITLE	Qty	Retail	Total
978-1-58197-638-0	WEST-B Basic Skills			
978-1-58197-609-0	WEST-E Biology 0235			
978-1-58197-565-9	WEST-E Chemistry 0245			
978-1-58197-566-6	WEST-E Designated World Language: French Sample Test 0173			
978-1-58197-557-4	WEST-E Designated World Language: Spanish 0191			
978-1-58197-614-4	WEST-E Elementary Education 0014			
978-1-58197-636-6	WEST-E English Language Arts 0041			
978-1-58197-634-2	WEST-E General Science 0435			
978-1-58197-637-3	WEST-E Health & Fitness 0856			
978-1-58197-635-9	WEST-E Library Media 0310			
978-1-58197-674-8	WEST-E Mathematics 0061			
978-1-58197-556-7	WEST-E Middle Level Humanities 0049, 0089			
978-1-58197-568-0	WEST-E Physics 0265			
978-1-58197-563-5	WEST-E Reading/Literacy 0300			
978-1-58197-697-7	WEST-E Social Studies 0081			
978-1-58197-639-7	WEST-E Special Education 0353			
978-1-58197-633-5	WEST-E Visual Arts Sample Test 0133			
	SUBTOTAL		**Ship**	$8.25
	FOR PRODUCT PRICES VISIT WWW.XAMONLINE.COM		**TOTAL**	